IET ENERGY ENGINEERING SERIES 201

Design, Control and Monitoring of Tidal Stream Turbine Systems

Other volumes in this series:

Design, Control and Monitoring of Tidal Stream Turbine Systems

Edited by
Mohamed Benbouzid

The Institution of Engineering and Technology

Published by The Institution of Engineering and Technology, London, United Kingdom

The Institution of Engineering and Technology is registered as a Charity in England & Wales (no. 211014) and Scotland (no. SC038698).

First published 2023

The Institution of Engineering and Technology
Futures Place
Kings Way, Stevenage
Herts SG1 2UA, United Kingdom

www.theiet.org

British Library Cataloguing in Publication Data
A catalogue record for this product is available from the British Library

ISBN 978-1-83953-420-1 (hardback)
ISBN 978-1-83953-421-8 (PDF)

Typeset in India by MPS Limited

Cover image: Shaunl/E+ via Getty Images

Contents

About the editor

Mohamed Benbouzid is a professor of electrical engineering at the University of Brest, France. His research interests and experience include analysis, design, and control of electric machines, variable-speed drives for traction, propulsion, and renewable energy applications, and fault diagnosis of electric machines. He is an IEEE fellow and an IET fellow, serving as an editor-in-chief of the *International Journal on Energy Conversion* and the *Applied Sciences (MDPI) Section of Electrical, Electronics and Communications Engineering.*

Introduction

Mohamed Benbouzid[1] and Demba Diallo[2]

The worldwide potential of electric power generation from tidal currents is enormous. The high load factor resulting from the fluid properties and the predictable resource characteristics make tidal energy resources attractive, and advantageous for power generation when compared to other renewable energies. Technologies are just beginning to reach technical and economic viability to make them potential commercial power sources shortly. While only a few small projects currently exist, the technology is advancing rapidly and has huge potential for generating bulk power. Moreover, international treaties related to climate control and the depletion of fossil fuel resources provide incentives to exploit energy from marine renewable sources. Several demonstration projects have been planned to harness tidal energy, a number of which have now reached maturity and could be deployed. However, the academic world knows very little about these technologies beyond the basic principles of their energy conversion. While research is more focused on hydrodynamics and turbine design, there is limited activity on the power conversion interface (including the electric generator), control, monitoring, and maintenance issues.

Due to the technological similarities between wind and tidal stream turbines, many problems encountered in tidal turbines have already been solved by wind turbine developers. Therefore, the knowledge and know-how gained from wind turbines have been transferred to accelerate the development of tidal stream turbine technologies. However, some fundamental differences in the design and operation of tidal stream turbines, such as biofouling and ocean current turbulence, need to be studied in more detail.

In competitiveness and large deployment contexts, this book, mainly research-oriented, addresses tidal stream turbines specific challenges; namely selection of the drivetrain configuration, and optimal control. Indeed, due to the submerged nature of tidal turbines, there is a potential higher failure rate. Consequently, there is a need to adopt the most resilient and maintainable drivetrain options, and develop resilient or fault tolerant control to increase the availability. In addition, biofouling must also be quantified and monitored to assess its influence on the hydrodynamic and electrical parameters of a tidal turbine. This book is primarily intended for researchers and postgraduates in the field of design, control, and monitoring of marine renewable energy harvesting systems. It provides methodologies and algorithms with several illustrative

[1]University of Brest, CNRS, Institut de Recherche Dupuy de Lôme, France
[2]Université Paris-Saclay, CentraleSupelec, CNRS, Group of Electrical Engineering Paris, France

examples and practical case studies. It includes extensive application features not found in academic textbooks. This book can also serve as a guide for prospective tidal stream turbine developers considering the main constraints related to the marine context: reliability, efficiency, and cost.

Chapter 1 deals with the presentation of design options for generators and drivetrains, which can be used for tidal stream generation to reach the challenging requirements of tidal energy generation in terms of reliability, efficiency, and system integration. It explores geared and direct-drive options (including rim-driven option), several possible technological options for the generators (induction generators, synchronous generators, doubly-fed induction generators, and unconventional generators). The chapter also discusses the technological options for gearbox including classical and specific magnetic gearboxes. Consequences of generators and drivetrains choices when using a specific optimal harnessing strategy associated with several turbine technological options will be discussed.

Chapter 2 is devoted to the development of control strategies for tidal stream turbines. Indeed, due to the submerged and/or semi-submerged operation, tidal stream turbines face several control challenges. First, an uncertain marine environment can introduce large amplitude external disturbances. Second, the control of the power electronic converters used to export the produced electrical power to the grid is challenging, as it requires accurate knowledge of the system parameters to guarantee stability and dynamic performance. However, marine environments can cause significant degradation to the system, resulting in rapid changes in these parameters. Finally, tidal stream turbines are often located in remote areas where the grid may be weak, i.e., the grid may have low inertia, which makes it more difficult to meet the grid code requirements. This chapter provides an overview of tidal stream turbine control methods that address these challenges.

Chapter 3 deals with the critical issue of tidal stream turbine resilience or fault–tolerant control. This chapter provides fundamentals on faults' resilience, which can be implemented at the design stage by using multiphase generators, redundancy in power converters, redundant sensors, cables, embedded control systems, and communication networks. Fault–tolerant control can be implemented with no additional hardware usage considering active and/or passive approaches. Tidal stream turbines fault resilience or fault–tolerant control is thereafter discussed based on a case study.

Chapter 4 addresses the condition monitoring of tidal stream turbines. Indeed, condition monitoring for fault detection and diagnosis is a challenging and critically important task due to tidal stream turbines' location and operating conditions. In this regard, there is a clear need for high reliability, given the severe limitations of maintenance access. This chapter, therefore, provides an overview of fault detection and diagnosis strategies for tidal stream turbines by addressing the specific fault of imbalance mainly generated by biofouling attachments.

Chapter 5 deals with the specific issue of biofouling in tidal stream turbines. It describes the biofouling development process and characterization and proposes a specific approach to assess the effect of attachments on the speed and generated power of a tidal stream turbine. The proposed approach, correlated to wind turbine icing phenomena, should be useful for biofouling detection and estimation.

Chapter 1

Tidal stream turbine generator and drivetrain design options

Jean Frédéric Charpentier[1], Khalil Touimi[2] and Mohamed Benbouzid[3]

1.1 Possible TST drivetrain options

1.1.1 Behavior of a tidal turbine in terms of torque/speed

The mechanical power which can be extracted from a horizontal axis tidal stream turbine can be calculated by the following equation:

$$P = \frac{1}{2}\rho C_p A V^3 \tag{1.1}$$

In this equation, sea water density ρ and turbine blade swept area A are considered as constants; V represents the tidal stream velocity; C_p is the power capture coefficient which depends on the operating point of the turbine (at first order, this power coefficient depends of the tip speed ratio, TSR, for a given value of the blade pitch). TSR is the ratio between the peripheral speed $R\Omega$ (where Ω is the turbine rotation speed in rad/s and R is the turbine external radius in meters) of the blades and the fluid velocity (V_{fluid} in m/s) incoming in the turbine disk (TSR = $\lambda = R\Omega/V_{fluid}$). For typical TSTs, the optimal value of C_p is estimated to be in the range of 0.35–0.5 (as for wind turbines). An example of C_p versus TSR curve is given in Figure 1.1.

Considering typical rated tidal stream velocity of about 3–4 m/s and (1.1), the required value of turbine external radius to reach a power in the MW range is about 10–20 m. As an example, the TST Sabella D10 from the French company Sabella which has been tested near Ushant Island in French Brittany which is shown in Figure 1.2 is characterized by a diameter of 10 m for a rated power of 1 MW.

For tidal systems, hydrodynamics phenomena as cavitation phenomenon limit the peripheral speed of the blades. Taking into account these considerations, MW range TST-rated rotational speeds are typically in the range of 10–30 rpm. Several

[1]French Naval Academy, Institut de Recherche de l'Ecole Navale, France
[2]Ecole Militaire Polytechnique, Actuators and Electromagnetic Devices Laboratory, Algeria
[3]University of Brest, CNRS, Institut de Recherche Dupuy de Lôme, France

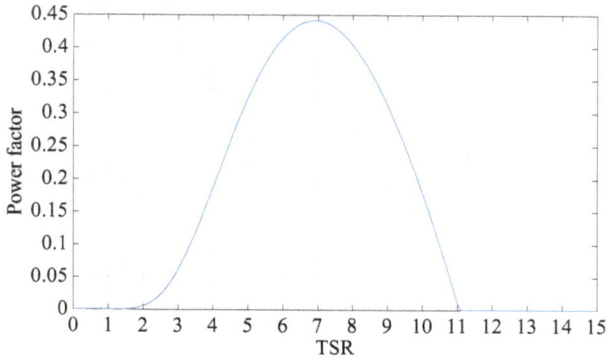

Figure 1.1   *Typical turbine power factor curve*

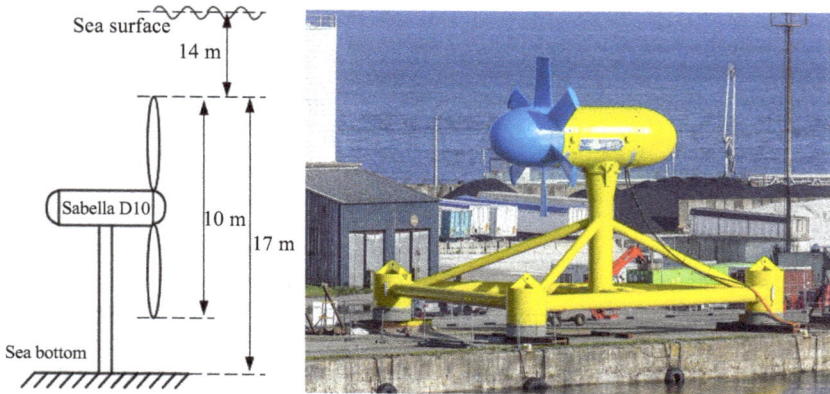

Figure 1.2   *Illustration and dimensions of Sabella D10 TST which have been tested in Ushant Island in France (courtesy of Sabella Company)*

possibilities of drive train can be used to connect mechanically the TST to its electrical generator.

An electrical machine in its rated operating point is characterized by the tangential stress created by the magnetic field sources in the rotor surface. This tangential stress which creates the machine torque as shown in Figure 1.3 is proportional to the product of the field densities created by stator and rotor in the air gap [1] and depends on the type of generator, the used materials (magnets, iron, coils) and the cooling technology.

Considering Figure 1.3, the torque T can be expressed as a function of the tangential stress ∂F_t and the rotor dimension (L, R, and V_{rotor} are the rotor length, radius and volume, respectively):

$$T = \partial F_t(2\pi RL)R = 2\partial F_t(\pi R^2 L) = 2\partial F_t V_{rotor} \tag{1.2}$$

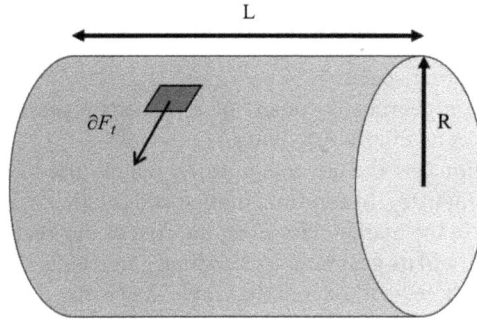

Figure 1.3   Schematic representation of the tangential stress created in the rotor surface of an electrical machine

On the one hand, the turbine can be connected directly to the generator (direct drive). In this particular case, generator specifications are characterized by a very low speed and very high torque (MN.m range). Because the rated torque of an electrical machine is directly linked by (1.2) to the machine volume, a direct-drive generator is obviously unconventional and is very huge in terms of dimensions. On the other hand, another option is to use a gearbox that allows increasing the generator-rated speed and consequently decreasing its rated torque (and generator volume) for a given level of power. It allows using a more conventional generator (higher speed, lower torque) but the gearbox option can be source of failure and it needs maintenance. Even though it can be noticed that gearbox reliability has been strongly improved in the last ten years [2–4]. Several possibilities in terms of generator and turbine association are detailed in the next paragraph.

1.1.2   Drivetrain options and generator configurations

Tidal turbines are subject to a very high loading torque comparing to wind turbines due to the marine current turbulence. Such environment harms the reliability of the TST especially the mechanical parts as the drivetrain subsystem. Indeed, various prototypes have been developed in the last decades where the main objective is to make such system more cost-effective comparing to the other renewable energy production systems. The choice of the TST drivetrain configuration is one of the important keys to enhance the availability and decrease the cost of produced energy. According to the drivetrain configuration, TSTs can be divided into three main groups: mechanical gearbox-driven TSTs where medium- to high-speed generators are employed, direct-drive TSTs with their low-speed high torque generators, and magnetically geared TSTs that are still under development [3].

1.1.2.1   Three-stage gearbox-driven tidal turbines

As explained previously, the more the generator operating torque is reduced the more compact it is. Whereas, when a low torque and high-speed rotating machine is chosen, a gearbox with a high conversion ratio is required. Consequently, the three-stage gearbox presents a suitable choice to match the turbine shaft low rotating speed with the generator high rotating speed.

Typical three-stage gearboxes consist of a planetary gearbox followed by two parallel axes gearboxes as Figure 1.4 shows.

Such configuration has the advantage to be widely used in other applications such as wind turbine industry. In addition, high-speed rotating machines are compact, robust, and available in the market. However, the cost of the gearbox depends mainly on its operating torque and its gear ratio. Accordingly, the three-stage gearbox is costly compared to the two-stage gearbox and the single stage-one.

In terms of reliability, three-stage gearbox failures causing replacement are low compared to the other TST components. However, as the access to the tidal turbine is difficult, the gearbox regular maintenance decreases the system availability and accordingly the cost of energy increases. For this reason, solutions as using platforms where the tidal turbine can be elevated for maintenance can be employed to mitigate the access difficulty. In [5], it was reported that during the testing period of the SeaFlow TST, the only major repair was due to a problem in the three-stage gearbox where some internal damage was revealed.

Concerning the efficiency, the more the number of gearbox stages get higher the more the power losses increases. Indeed, the three-stage gearbox efficiency of the SeaGen tidal turbine is around 95% [6]. On the other hand, most of three-stage gearbox-driven TSTs are based on induction generators which have a relatively lower efficiency when compared to permanent magnet synchronous machines. However, the use of power converters can increase the part-load efficiency.

Such drivetrain option is adopted by ANDRITZ HYDRO Hammerfest to develop the HS300 TST prototype with a rated power of 300 kW and the HS1000 pre-commercial TST with a rated power of 1 MW (Figure 1.5) [7].

Besides, Marine Current Turbines (MCT) realized the SeaFlow TST with a power rating of 300 kW (Figure 1.6a), the SeaGen 1.2 MW TST with its twin turbine, and the twin turbine SeaGen-S with a power rating of 2 MW (Figure 1.6b). In the

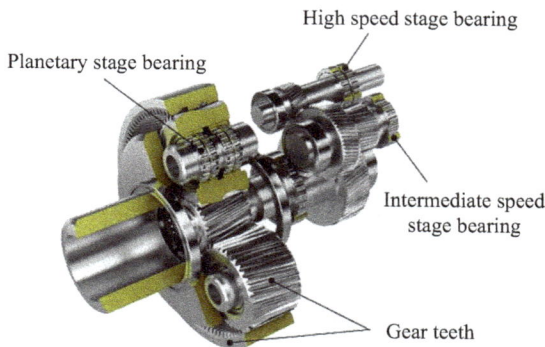

Figure 1.4   Typical three-stage gearbox

(a)   (b)

*Figure 1.5   ANDRITZ HYDRO Hammerfest tidal turbines [7]. (a) Illustration of the
             HS300 tidal turbine. (b) Illustration of the HS1000 tidal turbine.*

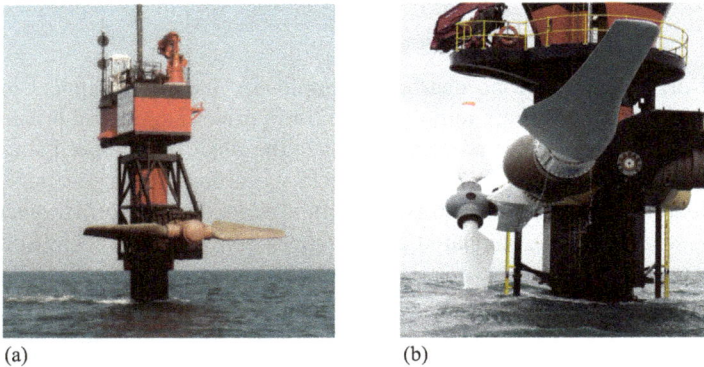

(a)   (b)

*Figure 1.6   Illustration of the TSTs developed by MCT. (a) Illustration of the
             SeaFlow tidal stream turbine [5]. (b) Illustration of the SeaGen tidal
             stream turbine [8].*

above cited projects, the three-stage gearbox is employed with a high-speed induction generator [5,7,8].

1.1.2.2   Two- and single-stage gearbox-driven tidal turbines

Contrary to the three-stage gearbox, in the single-stage and two-stage ones, the high-speed stage is removed. Consequently, the gearbox becomes compact, more efficient, and more reliable. However, its gear ratio becomes lower, which results in a larger generator as its operating torque becomes greater [9]. Such concept can be seen as a compromise between the conventional three-stage gearbox-based drivetrain and

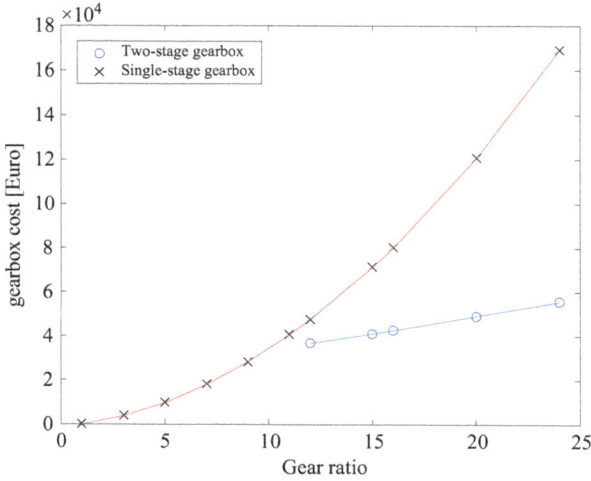

Figure 1.7   Single and two-stage gearbox estimated cost [9]

the direct-drive one where no gearbox is employed. Indeed, designers give other appellations to such drivetrain option as: multibrid drivetrain, hybrid drivetrain, and semi-direct drive one. In this context, to reduce the risk of high-speed stages of the gearbox in the wind turbine, a two-stage planetary gearbox-driven generator is developed by Alstom for the 5 MW M5000 wind turbine under the Multibrid project [10,11]. Moreover, several studies compared the drivetrain options of the wind turbines and they indicate that the gearbox-driven permanent magnet generators coupled with a single-stage or two-stage gearboxes present one of the promising solutions for more reliable and available systems [12–14]. Further, a comparison between the multibrid concept and the direct-drive one is performed in [9]. The study estimates the active material cost of single-stage planetary gearbox-driven permanent magnet generators with different gear ratios in comparison with the direct-drive configuration. The results show that the estimated cost of the planetary gearbox has a quadratic dependence on the gear ratio for a given operating torque (Figure 1.7). Moreover, the use of a single-stage planetary gearbox-driven generator can decrease the active materials cost of the generator and the gearbox by 40% when compared to a direct-drive one. Concerning the two-stage gearbox, which consists of a planetary gearbox coupled with a parallel shaft one, it is recommended for high gear ratios as depicted in Figure 1.7.

Otherwise, different tidal turbine prototypes with such drivetrain configuration are realized and tested in the last 15 years in many regions in China where it is shown that the semi-direct drive technology is highly reliable and efficient [15].

Some of the realized TST prototypes and precommercial projects with the discussed drivetrain are listed below:

- TST prototype with a rated power of 20 kW realized under the Science and Technology Development Project (STDP) [16].

(a)

(b)

(c)

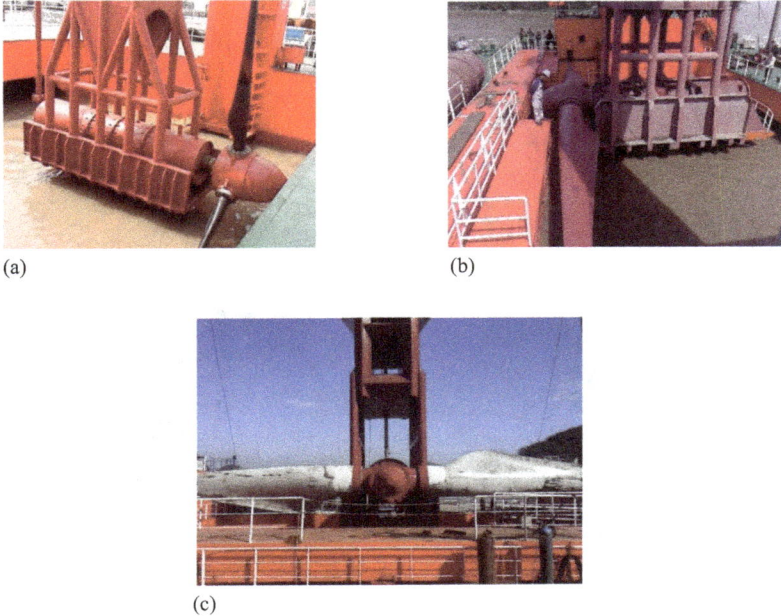

Figure 1.8   Illustration of semi-direct drive TSTs developed in China [15].
(a) Illustration of the semi-direct drive 60 kW tidal turbine.
(b) Illustration of the semi-direct drive 300 kW tidal turbine.
(c) Illustration of the semi-direct drive 650 kW tidal turbine.

- A TST with a rated power of 60 kW funded by the Chinese Marine Renewable Energy Special Fund (Figure 1.8(a)).
- ZJU in co-operation with GUPC (Guodian United Power Corporation) to develop a 300 kW TST (Figure 1.8(b)).
- Zhejiang University (ZJU) developed a 650 kW TST (Figure 1.8(c)).
- ATLANTIS developed a 1.5 MW commercial TST which presents the largest TST with such drivetrain option (Figure 1.9) [17].

1.1.2.3   Direct-drive tidal turbines

Since the beginning of the 1990s, direct-drive wind turbines have been designed to avoid the drawbacks of the mechanical gearbox and make the wind turbine drivetrain simpler and more reliable. Since then, many manufacturers have chosen the direct-drive option to build their wind turbines (e.g. Enercon, siemens, etc.) [18,19]. However, as the gearbox is removed, the turbine shaft torque is not reduced, which results in large diameter multi-pole generators (non-classical) as shown in Figure 1.10. Such generators, for the same power rating, are obviously more expensive if compared to high-speed generators.

Figure 1.9   Illustration of the commercial AR1500 tidal turbine [17]

Besides, with the blow up of research on TST systems in the last two decades, the direct-drive concept has been highly recommended to overcome the mechanical gearbox issues and its regular maintenance requirement. During this period, several industrial projects and prototypes of direct-drive TSTs have been developed [20–25], where permanent magnet-based generators are used. The low-speed high torque generators with their high diameter and low iron core thickness make possible to integrate the rotor of the generator at the periphery of the blades. Such concept, also known as rim-driven one, has been adopted for the Naval Group Open Hydro Tidal turbine [26,27].

In terms of reliability, the direct-drive permanent magnet generator is highly reliable due to the brushless rotor and the simplicity of the drivetrain. However, direct-drive wounded rotor synchronous generators are not recommended for tidal turbine applications, especially for high-power range turbines [3]. Besides, as the direct drive generators are low-speed machines, their cooling is easier than high-speed ones due to their structure. Moreover, the direct-drive TST has a high efficiency as the gearbox losses are removed.

To reduce the operating torque of the direct-drive TSTs, and consequently the volume and weight of such generators, contra-rotating turbines can be used where both the rotor and the stator rotate in opposite directions. Such a solution makes the operating torque of the machine reduced to half compared to a usual direct-drive generator as the relative rotational speed is doubled. In addition, in such a design, the reaction torque on the support structure is almost suppressed which reduce significantly its cost [28]. This concept is adopted by Nautricity Ltd to design the CoRMat tidal turbine with a power rating of 500 kW [29,30] (see Figure 1.11).

Figure 1.10   Large diameter multi-pole synchronous generator, Enercon E-126 (the rated operating point is 7.5 MW at 1 rpm) [17]

The main direct-drive TST projects, in addition to the CoRMat TST, are listed below:

- The Race Rocks demonstration project with its bi-directional ducted TST (Figure 1.12(a)).
- The rim-driven 250 kW prototype developed by OpenHydro [32] (Figure 1.12(b)).
- The rim-driven 2 MW TSTs developed by OpenHydro/Naval Group (Figure 1.12(c)).
- The HyTide 1000 TST developed by Voith Hydro with a 1 MW power rating (Figure 1.12(d)).
- The D10 projects developed by Sabella with a nominal power of 1 MW (Figure 1.12(e)).

1.1.2.4   Magnetically geared tidal turbine

Gears are used to increase the torque density of the TST. Despite the development of mechanical gears technology, the gearbox remains being a critical component with a

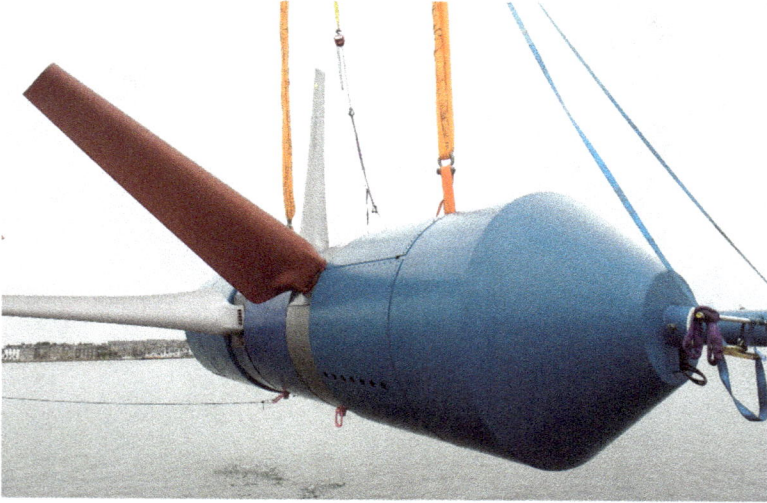

Figure 1.11   Illustration of the 500 kW contra-rotating tidal turbine CoRMat [31]

high risk priority number (RPN) [4]. Besides that, the modern magnetic gears, with the development of rare earth permanent magnets, are achieving high torque densities which make them competitive to mechanical ones [37–43]. The magnetic gears are analogous to mechanical gears; however, torque transmission is done by means of attraction and repulsion forces between the rotating magnetic parts. As the torque transmission is contactless, the turbine shaft and the generator shaft are physically isolated. Therefore, the magnetic gear is advantageous as it does not require any lubrication and it is tolerant inherently to overload. Furthermore, the non-existence of mechanical gearing reduces vibrations and acoustic noise emission, it increases the efficiency of the system, and extends its lifetime. All these advantages are at the expense of a higher capital cost, but with a reduced operation and maintenance cost.

Magnetic gears have different topologies which can be divided into two main groups, the first one includes magnetic gears inspired by mechanical ones, such as parallel axis magnetic gears, planetary magnetic gears, cycloidal magnetic gears, and strain wave magnetic gears also known as harmonic magnetic gears. In these topologies, gear teeth are replaced by permanent magnets, hence a limited number of poles contribute to the torque transmission in each instant as shown is Figure 1.13.

On the other hand, the second group consists of flux-modulated magnetic gears where all the poles participate in the torque transmission process instantaneously. Figure 1.14 shows a typical flux-modulated permanent magnet gear which presents the preferred topology as most of the recent publications on this field focus on it.

To ensure a coupling between the outer rotor and the inner one which have different number of poles (p_{ou} and p_{in}), a modulator with Q_m ferromagnetic pole pieces (segments) is inserted between them. The modulator creates additional space harmonics of the magnetic flux density coming from the inner rotor which have an order of

(a)   (b)

(c)

(d)   (e)

Figure 1.12   *Illustration of the main direct-drive tidal turbine projects. (a) The*
               Race Rocks demonstration TST [33]. (b) The 250 kW rim-driven Open
               Hydro prototype [32].(c) The 2 MW rim-driven Open Hydro TST [34].
               (d) Illustration of the HyTide 1000 TST [35]. (e) Illustration of the
               Sabella D10 TST [36].

$Q_m + p_{in}$ and $Q_m - p_{in}$. The modulator creates also space harmonics of the magnetic
flux density coming from the outer rotor where the modulated flux density has an
order of $Q_m + p_{ou}$ and $Q_m - p_{ou}$. Therefore, to have a magnetic coupling between the
inner and the outer rotors, only two configurations are possible as shown in (1.3):

$$Q_m = p_{ou} \pm p_{in} \tag{1.3}$$

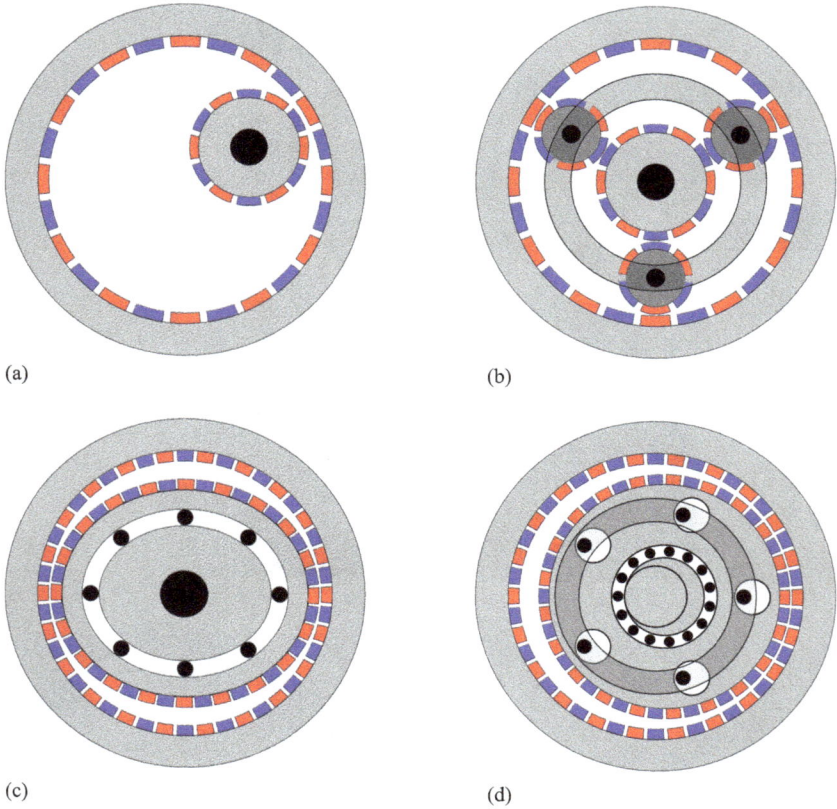

Figure 1.13   Magnetic gears. (a) Parallel axis magnetic gear. (b) Planetary magnetic gear. (c) Strain wave magnetic gear. (d) Cycloidal magnetic gear.

Further, the flux-modulated magnetic gear can work as a power splitter if the three rotors are not blocked, (1.4) shows the relation between the rotational speed of the three rotors in such case:

$$Q_m \Omega_s = p_{ou} \Omega_{ou} \pm p_{in} \Omega_{in} \tag{1.4}$$

where Ω_s, Ω_{ou}, and Ω_{in} are the modulator, inner rotor, and outer rotor angular speeds, respectively.

To have an operating mode of a gearbox, the outer rotor or the modulator is in general kept stationary. Equations (1.5) and (1.6) provide the gear ratio in the two cases:

$$r_{io} = \frac{\Omega_{in}}{\Omega_{ou}} = \mp \frac{p_{ou}}{p_{in}} \tag{1.5}$$

$$r_{is} = \frac{\Omega_{in}}{\Omega_s} = \pm \frac{Q_m}{p_{in}} = 1 \pm \frac{p_{ou}}{p_{in}} \tag{1.6}$$

Outer rotor back iron
Outer rotor PMs
Modulator
(pole pieces)
Inner rotor PMs
Inner rotor back iron

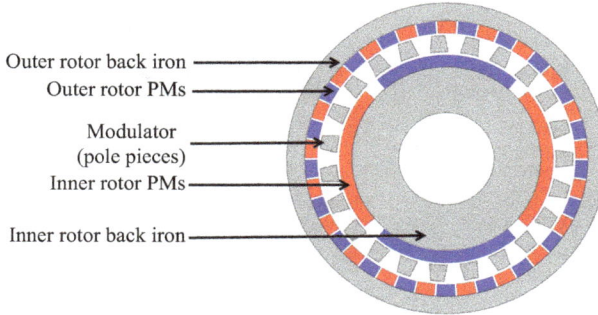

Figure 1.14   Cross-sectional view of a flux-modulated magnetic gear

where r_{io} is the gear ratio when the modulator is stationary and r_{is} is the gear ratio when the outer rotor is fixed.

The flux-modulated magnetic gear with its advantages can be integrated with an electrical generator thanks to the electromagnetic properties of the magnetic gear. The magnetically geared generators are composed generally of four components, the stator, the outer rotor, the modulator, and the inner rotor. The stator can be integrated with the modulator to have a wound-modulator magnetically geared machine. In addition, it can be integrated with the outer rotor which gives the pseudo-direct-drive machine structure. Meanwhile, the stator can replace the inner rotor or it can be inserted inside it. In such case, the generator and the magnetic gear share the same high-speed inner rotor. Such a design has three airgaps unlike the previously cited designs with only two airgaps. Otherwise, a generator structure with a single airgap known as the Vernier generator, despite its poor power factor, presents an interesting alternative due to its simplicity. In such a structure, the stator tooth tips modulate the magnetic flux (Q_m pole pieces) coming from the outer rotor permanent magnets (p_{ou} pole pairs). The magnetic flux is canalized through the stator teeth (Q_s teeth) to induce an electromotive force in the stator with its p_{in} pole pairs winding. Figure 1.15 illustrates a typical outer rotor Vernier generator where the stator teeth number is equal to the modulator pole pieces ($Q_s = Q_m$).

The flux-modulated magnetically geared generators with their multiple rotors give the possibility of using a contra-rotating turbine, where the input high torque comes from both the outer rotor and the modulator rotating in opposite directions. Such a configuration has the advantage of being more compact as it offers a higher equivalent gear ratio (almost double) without changing the number of poles and the machine structure [39].

The flux-modulated magnetically geared generators with its different structures have a higher torque density than direct-drive multi-pole TSTs and they are more reliable than the mechanically geared ones. Despite the multiple advantageous that can offer such drivetrain technology, it is still being under development and it is not mature enough to be widely deployed in industrial TST projects.

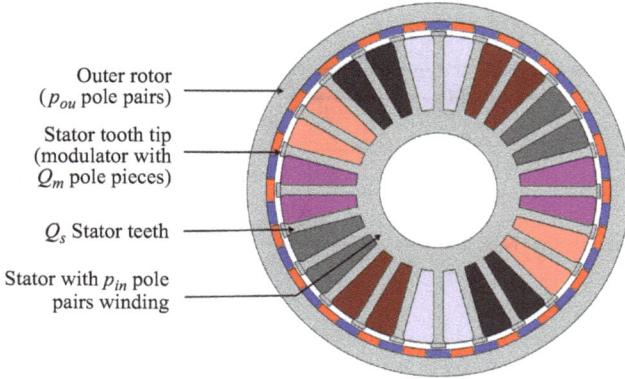

Outer rotor
(p_{ou} pole pairs)

Stator tooth tip
(modulator with
Q_m pole pieces)

Q_s Stator teeth

Stator with p_{in} pole
pairs winding

Figure 1.15   *Illustration of a typical Vernier generator structure*

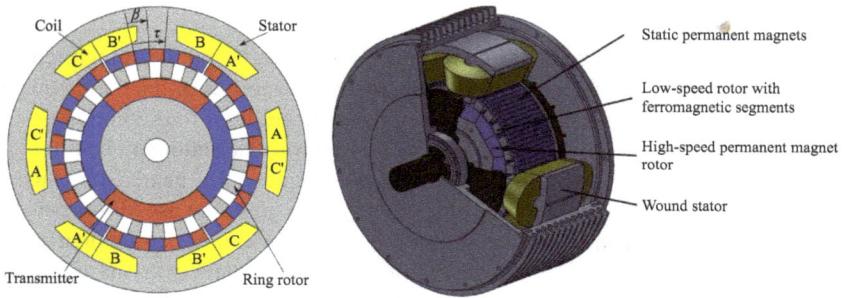

Coil   Stator

Transmitter   Ring rotor

Static permanent magnets

Low-speed rotor with
ferromagnetic segments

High-speed permanent magnet
rotor

Wound stator

Figure 1.16   *Illustration of the pseudo-direct drive structure [46]*

In this context, few prototypes are realized with the same concept for TST applications. Magnomatics and Seaplace developed a Pseudo-Direct Drive (PDD) generator for a floating TST with a power rating of 6 kW (see Figures 1.16 and 1.17) [44,45].

In this topology, the embedding is done between the outer rotor of the magnetic gear and the stator of the generator which are both stationary. The modulator is connected to the turbine shaft.

Indeed, the PDD concept has been proposed by Atallah *et al.* [47] with a measured torque density in excess of 60 kN.m/m^3 which is much higher than the torque density of a standard machine (around 10 kN.m/m^3). The same concept is proposed for wind power generation by Penzkofer *et al.* [48].

Besides, a 500 kW PDD generator is realized basing on the magnomatic's PDD design (Figure 1.18). According to the case study in [49], the performance and endurance testing shows that the generator could deliver power with high efficiency which could reduce the Levelized Cost of Energy (LCOE) of around 3%.

In addition to the PDD generators, several prototypes are realized for wind turbine applications and can be mirrored for the use in TST projects. In [50], the authors

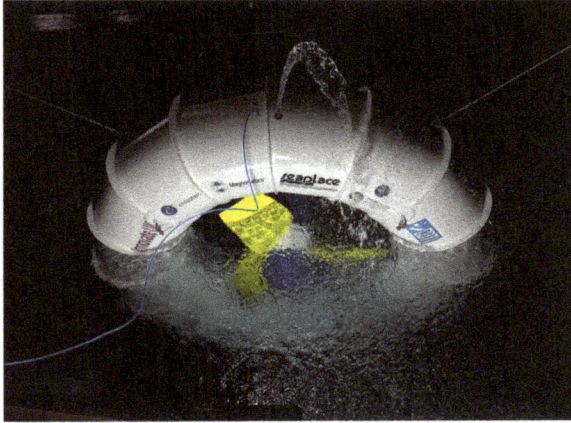

Figure 1.17   The floating PDD TST prototype (©Magnomatics) [44]

Figure 1.18   Illustration of the 500 kW pseudo-direct drive generator
(©Magnomatics) [49]

present a magnetically geared outer rotor permanent magnet generator. In this proto-type, the outer rotor of the permanent magnet generator is integrated with the magnetic gear inner rotor while the magnetic gear outer rotor is connected to the turbine shaft as depicted in Figure 1.19. The simulated torque density of the developed device is 87 kN m/m^3.

Axial flux configuration is also proposed for integrated magnetically geared generators. Johnson *et al.* [51] developed a prototype where the axial flux permanent

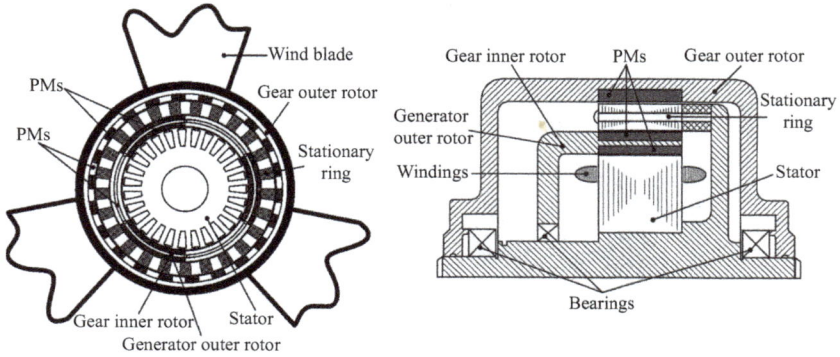

Figure 1.19   Illustration of a magnetically geared outer rotor permanent magnet generator [50]

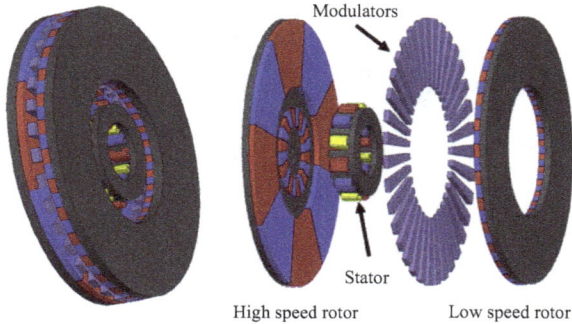

Figure 1.20   Illustration of the axial flux magnetically geared generator [51]

magnet generator is integrated in the radial bore of the axial flux magnetic gear (see Figure 1.20). The proposed design has a simulated torque density of 60.6 kN.m/m^3.

1.1.3   Power harnessing general strategy

It can be noticed that tidal resource is characterized by tidal current velocity high variability. Tidal velocity time vectors can be determined years in advance in given location thanks to the predictability of tidal phenomenon [52–54]. As an example, Figure 1.21 presents a time series of the tidal current velocity value during, 1 day, 1 month, and 8,424 h, respectively for one site located in the Raz de Sein, Brittany, France. The considered velocities are considered to be aligned along a common direction.

This assumption is in first order true in a large part of high potential tidal stream sites. This is why there are positive and negative values in the curves of Figure 1.21. In Figure 1.22, the corresponding occurrences distribution during the whole period

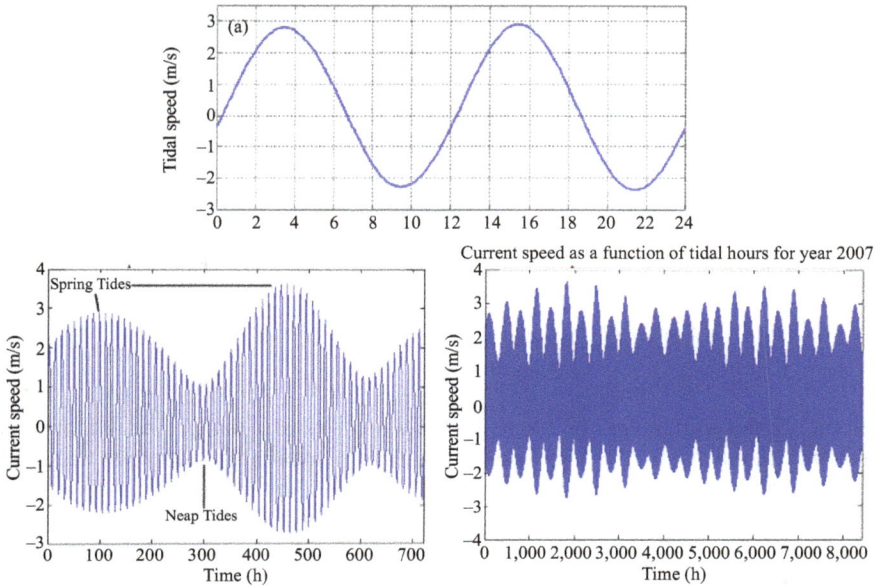

Figure 1.21   Typical time series of tidal current velocity in Raz de Sein (France) [52]

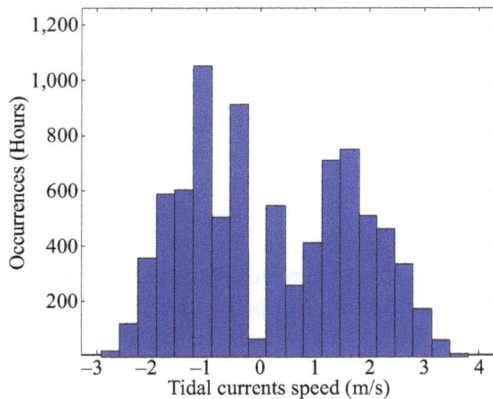

Figure 1.22   Occurrences distribution of tidal current velocity in Raz de Sein (France)

(8,424 h) corresponding to the time series of Figure 1.21 are shown. Considering this occurrence distribution, it can be noticed that only a very few number of hours corresponds to high velocities.

As an example considering the data of Figures 1.21 and 1.22, velocities greater than 2.5 m/s occur only during 7% of the total considered time. However, these high

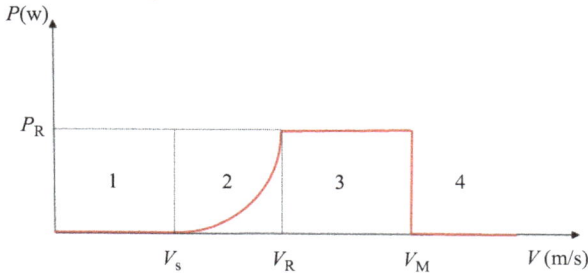

Figure 1.23   Power extraction strategy with four typical control areas

velocity values lead to very highly extracted power values if the turbine is controlled to operate at the maximal value of power coefficient (maximal power point tracking). As an example, the extracted power using MPPT for 12 m diameter turbines with an optimal power coefficient of 0.46 will be 416 kW and 1.27 MW for tidal speeds of 2.5 m/s and $V_{max} = 3.63$ m/s (a maximal value of current in Figure 1.22), respectively. To be able to find a compromise between extracted energy amount and system costs, power limitation strategy can be used for high value of tidal current velocity as done for MW range wind turbines. This strategy allows maximizing extracted energy for a limited capital expenditure cost (CAPEX). This power harnessing strategy is described in Figure 1.23.

The strategy corresponds to four main areas of control. In the first area where the tidal stream velocity is lower than a starting value V_S, the turbine is stopped because the extracted energy is not sufficient to compensate chain energy losses. In area 2, where the tidal stream velocity is included between the starting value V_S and the rated value V_R, the system is controlled to harness a maximal power (maximal power point tracking). In area 3 where the tidal stream is higher than the rated value V_R, the system is controlled to harness a fixed value of power which corresponds to the power extracted at the rated operating point V_R, $P_R = P_{LIM}$. A 4th area corresponds to very high value of tidal stream (greater than the maximal operation value V_M). In this 4th area, the system can be stopped and secured for safety reason. Thanks to the limited range of tidal current values, the 4th area is not always necessary to be used for the power management of a tidal stream system. In [55], it is shown that a power limitation of about only 30% of the maximal power ($0.3P_{max}$) obtained using MPPT strategy for maximal value of current velocity leads only to a reduction of 10% of the global energy harnessing. These results have been obtained using the resource data illustrated in Figures 1.21 and 1.22. In Figure 1.24, the rate of extracted power (in percent) versus power limitation level (in percent) is presented. E_t (100%) is the extracted energy without power limitation [only MPPT (zone 2) is considered to establish this reference value for all the tidal stream velocities]. The figure x-axis is the ratio in percent between the rated (or power limitation) $P_R = P_{LIM}$ and P_{MAX} ($100(P_R/P_{MAX})$) where P_{MAX} is the maximal power extracted at V_{MAX} using only MPPT strategy (supposing that the system is operated in zone 2 until V_{MAX}).

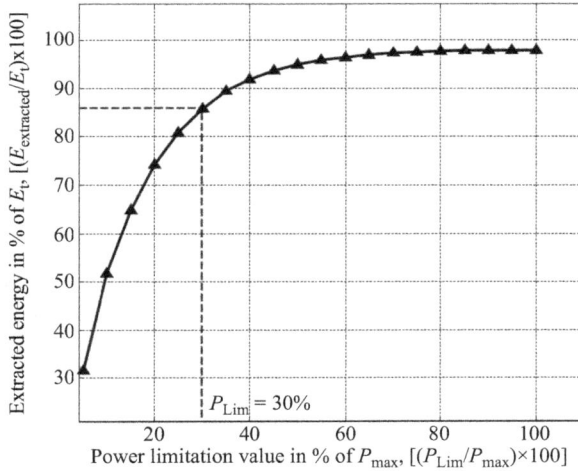

Figure 1.24   *Extracted energy (%) versus power limitation level from [55]*

These analyses show that MPPT combined with power limitation strategy leads to a strong reduction of energy chain sizing without a significant reduction of global energy harnessing. This is why this strategy is very attractive for tidal stream turbines power management. In the next section, the possible technical ways to follow this power extraction strategy will be examined.

1.1.4 *Turbine control options*

1.1.4.1 MPPT strategy

Maximizing the extracted power leads to maintain the turbine at its optimal operating point. For this purpose, the turbine rotating speed has to be controlled to maintain the TSR value at its optimal value which corresponds to the maximal power coefficient values. That means that for each tidal stream velocity, the turbine rotational speed will be controlled to an optimal value which is in a first order proportional to the tidal current velocity (TSR = optimal TSR). This can be done using an AC generator and drive association which allows generator (and coupled turbine) speed control and power injection to the grid. Several machine and drive configurations can be used for this purpose as for example squirrel cage induction generators, wound rotor synchronous generators, permanent magnet synchronous generators or doubly-fed induction generator, and their corresponding drives. Figure 1.25 describes the principle of generator and drive systems for a squirrel cage induction generator, doubly-fed induction generator, and permanent magnet synchronous generator (PMSG). Classically the AC converter is made of two different converters (grid side converter and generator side converter) which are connected via a DC bus. The DC bus voltage, V_{dc}, is controlled to a fixed value by the grid side converter (3-phase IGBT PWM Bridge)

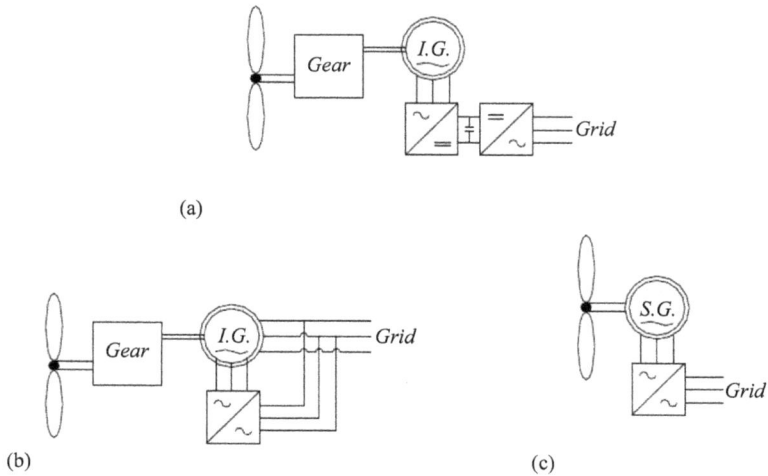

Figure 1.25   Examples of generators and drives association. (a) Geared squirrel
cage induction generator and drive. (b) Geared doubly-fed induction
generator and drive. (c) Direct-drive synchronous generator (PM or
wound rotor) and drive.

which insures a full transit of active power from the DC bus to the grid and the control
of injected reactive power in the grid. The generator side converter (which is also a
3-phase IGBT PWM bridge) is used to control the rotating speed of the generator by
controlling the generator braking torque using AC electrical machines variable speed
control method as the well-known flux-oriented control (FOC) methods. More details
on the control of these two converters can be found in [56,57]. This strategy is applied
in MPPT zone (zone 2 of Figure 1.23).

1.1.4.2   Power limitation using pitch control

A first solution to control the tidal turbine in power limitation area (zone 3) is to
use a pitch control system as in MW range Wind Turbines. Pitch control is an active
mechanical system which is located in the hub of the turbine. This system allows
modifying in real time the angle (pitch) between the blades and the hub axis. Principle
of pitch control and examples of pitch system mechanisms are depicted in Figure 1.26.

Acting in the pitch system allows changing the turbine geometry and modifying
the power coefficient characteristics of the turbine. As an example, the C_p curves of
the experimental turbine described in [58] are given as a function of Tip Speed Ratio
and pitch angle in Figure 1.27. These curves are obtained by numerical calculation
using blade element momentum theory method [52]. This theory is a useful tool for
modeling tidal turbine without complex CFD methods [59]. It can be noticed that
acting in the pitch angle allows: on the one hand, to limit the TSR range of the turbine
and, on the other hand, to decrease the power coefficient values. This is why acting

Figure 1.26   Illustration of pitch systems

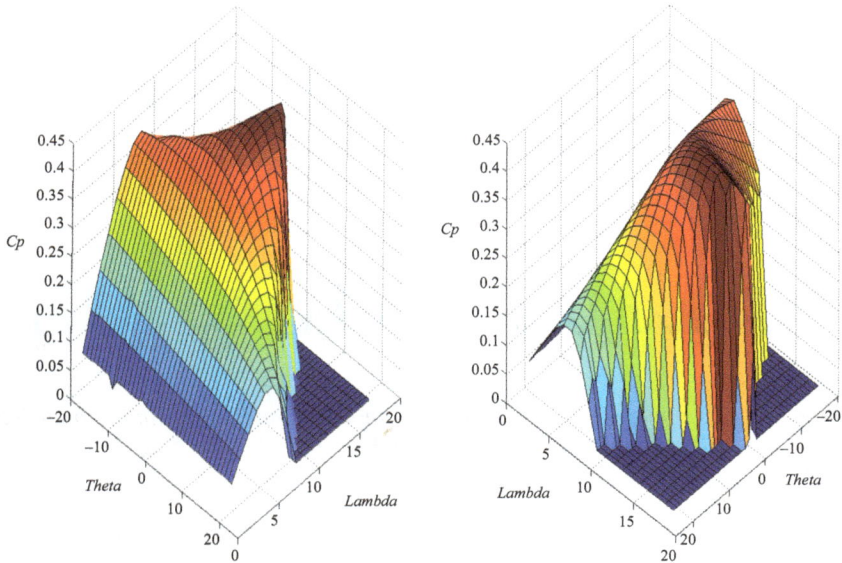

Figure 1.27   Power coefficient curves as a function of tip speed ratio (Lambda) and pitch value (Theta) for a variable pitch turbine[52]

in pitch control can allow limiting the extracted power and maintaining the rotating speed to its rated value.

In this case, the turbine rotating speed is controlled to a fixed value which corresponds to the maximal speed which is reached in MPPT area (rated rotating speed)

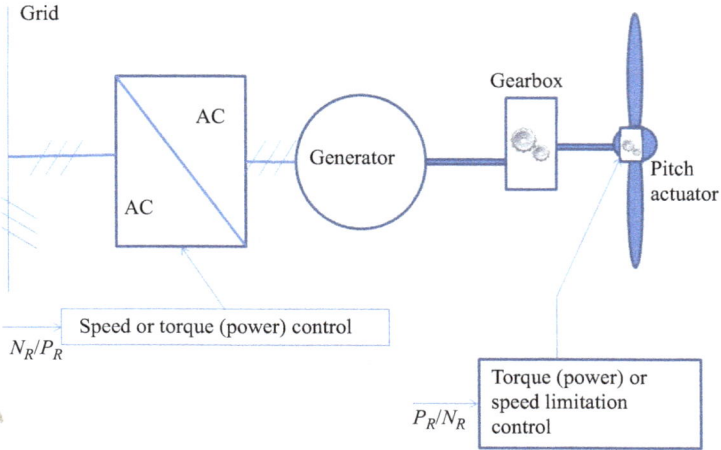

Figure 1.28   Principle of control in zone 3 (power limitation) for a variable pitch turbine

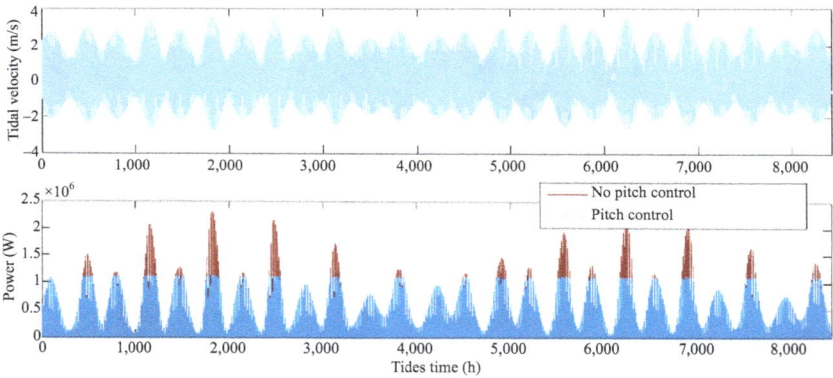

Figure 1.29   Example of extracted power (with MPPT and power limitation using pitch control) for a variable pitch turbine [52]

by the same way as in zone 2 (generator and drive control). The power is limited to its rated value (P_R) acting in the pitch control. Another possible solution is to use the pitch system to control the speed as constant and the generator drive to control the power at its rated value. The global scheme of turbine control in zone 3 for a variable pitch turbine is given in Figure 1.28.

As an example, this kind of control (MPPT and power limitation with pitch control) is simulated for the same tidal time series as in Figure 1.21 for a 1.2 MW rated power turbine. The extracted power is then given in Figure 1.29 and it can be noticed that the combination of the two actions (speed/torque control and pitch control) allows to follow efficiently the power harnessing strategy of Figure 1.23).

Using a pitch control system is in fact a very advantageous option in terms of control and turbine design: the turbine rotational speed range and the generator are both limited which allows limiting the constraints on the choice of generator and drive and their control. In fact, with a pitch control system, any kind of generator and drive can be used: induction generator, doubly fed induction generator, or synchronous generators with or without gearboxes.

1.1.4.3   Power limitation with fixed pitch turbines using underspeed or overspeed and consequence on the torque VS speed generator and drive characteristics

However, a variable pitch control system is a possible source of failure and requires regular maintenance. This is why in the particular context of tidal stream energy harnessing, using a fixed pitch turbine can be an interesting option in order to increase the global robustness of the whole system. This option has been considered by several companies for their prototype turbines [60]. As examples Voiht Hydro, OpenHydro and Sabella have tested large marine fixed pitch tidal current turbines. For the MPPT control strategy (area 2 of Figure 1.23), the control is similar to the control of a variable pitch turbine. For the third area (area 3 of Figure 1.23), the only possible way to limit the power is to use a particular speed control strategy in order to limit the harnessed power. In [55,61,62], several possible strategies are studied in order to control a fixed pitch tidal stream turbine in area 3. Figure 1.30 presents the power curves corresponding to several tidal stream speeds versus the turbine rotating speed for a fixed pitch turbine. These curves can be deduced, for a given turbine geometry, of the C_p curve of the turbine. In Figure 1.30, MPPT strategy is illustrated by the black curve trajectory until the turbine-rated point that corresponds to a rotating speed of $N_R = 22$ rpm for a harnessed power $P_R = 400$ kW in Figure 1.30. In order to limit the power, two possible options can be used when the tidal velocity will be greater than the rated value (corresponding to the rated point).

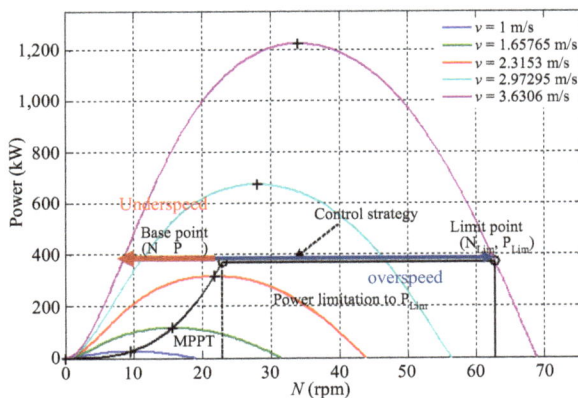

Figure 1.30   Power extraction strategy for a fixed pitch turbine: underspeed and overspeed power limitation

The first one is to use an underspeed strategy [61]. In this case, the turbine speed is controlled at a rotating speed smaller than the rated speed, N_R, in order to limit the power to the rated power. Then the operating point of the turbine for a given tidal speed velocity is the left-side intersection of the power curve (corresponding to the given velocity) and the $P = P_R$ horizontal curve. This strategy corresponds to the red part of the trajectory in Figure 1.30.

The second option is to use an overspeed strategy [55,62]. In this last case, for tidal velocities greater than the rated one, the turbine rotating speed will be controlled to reach the operating points which corresponds to the right-side intersections of the power curves (corresponding to the tidal velocity) and the $P = P_R$ horizontal curve. This strategy corresponds to the blue line trajectory in Figure 1.30.

These two strategies lead to different constraints in terms of generator control and design. In Figure 1.31 are presented the operating point trajectories corresponding to Figure 1.30 in the power versus rotating speed and torque versus rotating speed frames related to an underspeed power limitation strategy. Considering that for the area 3, the operating points correspond to a constant power operation ($P = P_R$). The corresponding torque characteristics leads to a hyperbolic curve $T = P_R/\Omega$. That means that the torques which will be reached for the maximal tidal stream velocities (point 2 in Figure 1.31) correspond to small rotating speeds for a fixed power (rated power). The maximal tidal stream velocities lead to operating point corresponding to very high torque value. In this case, the corresponding torque value can be two to three times higher than the torque achieved at the turbine rated point (point 1 in Figure 1.31: maximal operating point for MPPT area (area 2)). This maximal torque value will be directly linked to the shape of the C_p curve and the range of tidal speed velocities. High maximal tidal speed and steep slope of the first part of the C_p curve lead to increasing the maximal torque values. This is why this strategy will lead to

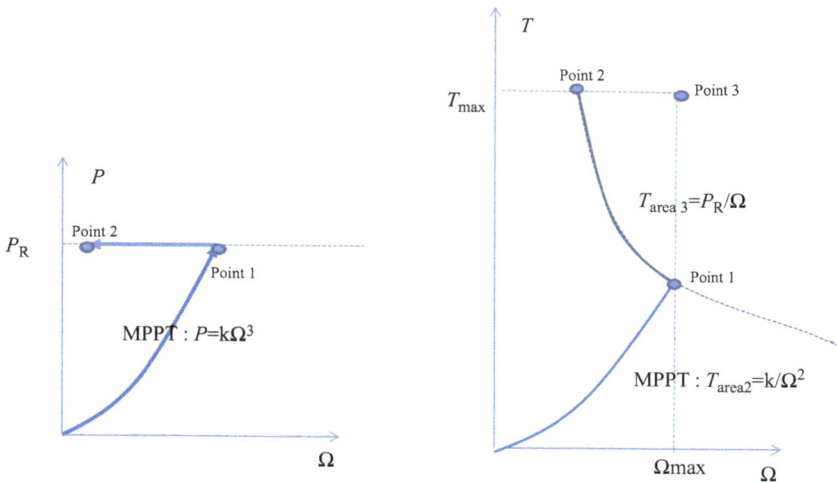

Figure 1.31   Underspeed power limitation: trajectories in power and torque VS speed

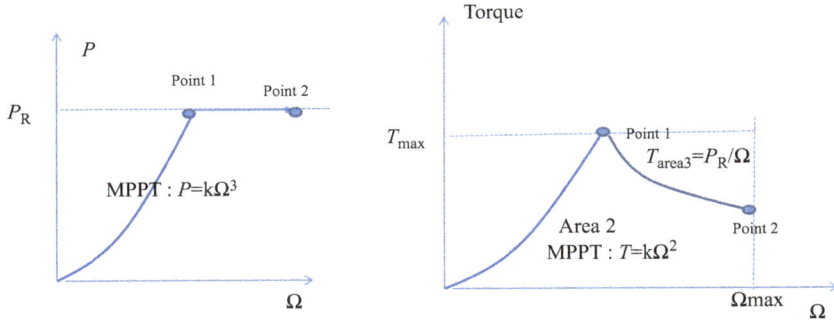

Figure 1.32 Overspeed power limitation: trajectories in power and torque VS speed

a strong over-sizing of the generator because, for a given technology, the volume of the generator is directly linked to the torque to be controlled by the system as shown in (1.2). Another important constraint is related to the control stability. It can be demonstrated that the first part of the power and torque curves of the turbine corresponds to unstable mechanical behavior of the system [61]. That means that turbine speed control is less easy for the operating points of area 3 in underspeed strategy than for the other operating points of the system.

Figure 1.32 presents the trajectories obtained for an overspeed strategy in the power and torque versus speed frames. As for underspeed strategy, the corresponding torque characteristics in area 3 leads to a hyperbolic curve $T = P_R/\Omega$ but the operating points correspond to higher rotating speeds than the rated speed (right part of the hyperbolic curve). In this case, this is why high values of the tidal velocity correspond to a strong decreasing of the torque that is to be controlled by the system. In this case, the generator global sizing corresponds to the maximal torque which is obtained for the rated point. On the contrary of the case of underspeed strategy, the corresponding operating points are mechanically stable which tends to facilitate the speed control of the turbine. However, it is obvious that high velocity leads to high turbine rotating speeds. The higher rotating speed values will be directly related to the velocity range and the slope of the right part of the C_p curve. Flat slope of the right part of the C_p curve leads to very high values of turbine rotating speeds [63]. Operating the system in these high rotating speeds can lead to very strong mechanical constraints on the turbine and high constraint on the design of generator and drive systems which have to be chosen in order to be operated in a very large range of speed at a constant power.

1.2 TST generator design

1.2.1 Design criteria

Considering the specific context of tidal stream energy harnessing, it is obvious that minimizing maintenance and increasing reliability of generator and drive system is a key feature because such systems are particularly difficult to access during the whole lifetime of the turbines. Another critical point is the generator and drivetrain compactness: MW range tidal stream turbines blade diameters are comparatively 3–4 smaller than MW range wind turbine ones for the same power but their rotational speeds are similar. That means that integrating generator and drivetrain in a nacelle in the axis of the turbine without impacting significantly the hydrodynamic behavior of the system can be a challenging point. Minimizing the global cost (OPEX and CAPEX) of the system is also a design goal. All these considerations lead to key criteria for the choice of drive train, generator, and drive association.

1.2.2   Possible generator type depending of the drivetrain options

As said in previous paragraphs, several kinds of AC generators and drives can be theoretically used to be associated with tidal stream turbines. A possible type of AC generator to be used depends of the drivetrain and pitch options. Using a doubly-fed induction generator (DFIG) which is a very popular option for MW range onshore wind turbines can be a track to reduce system CAPEX. In a DFIG, the stator windings are directly connected to the grid and the generator speed can be controlled using a reduced size (and cost) back to back power converter connected to rotor winding as shown in Figure 1.25(b) [57]. The converter control allows a field-oriented control of torque and speed of the DFIG as presented in [57]. Using DFIG allows reducing strongly the cost of power electronic devices because the converter can be sized to only a fraction of the rated power of the generator. In fact, the relative rating of the converter depends directly on the speed range in which the system is operated. The relation between the converter rated power and the maximal value of the slip is given by (1.7):

$$P_{conv} = \frac{s_{max}}{1 + s_{max}} P_R \qquad (1.7)$$

s_{max} is the maximal absolute value of the generator slip (which can be positive for low speeds and negative for high speeds). P_{conv} and P_R are the converter-rated power and the turbine-rated power, respectively.

Table 1.1 gives possible speed ranges and converter sizing ratio. In this table, K is the ratio between the maximal and minimal generator rotating speeds (respectively, $N_{max} = N_s(1 + s_{max})$ and $N_{min} = N_s(1 - s_{max})$) which can be managed by the system where N_s is the synchronous speed of the generator. $Rc = P_{conv}/P_R$ is the ratio between the converter-rated power and the turbine-rated power.

It is then obvious considering previous paragraphs that using a fixed pitch turbine will lead to increase too much of the operating speed range of the turbine and generator set in the power limitation control area. This is why a pitch control system is needed in order to limit both power and rotating speed if a DFIG-based system is used. DFIG is also an induction generator and it is difficult to design and induction generator with very high pole number. Because increasing the number of pole allows minimizing the volume and mass of electrical machine iron core [1], electrical machines with low number of poles are less compact than high number ones for the same specifications in terms of torque. The tangential stress in the rotor (or stator surface) which creates

Table 1.1   Typical sizing ratio for the converter for several speed ranges for a DFIG

$K = N_{max}/N_{min}$	s_{max}	$R_c = P_{conv}/P_R$
2	0.33	0.25
3	0.5	0.33
4	0.6	0.375

the machine torque (1.2) is also lower in a DFIG than in synchronous machines for a given cooling technology. These two assumptions limit the possible compactness of high torque low speed DFIG. This is why a DFIG generator has to be associated with a gearbox to be integrated with a tidal turbine. One can notice that in some works on wind turbine, a single-stage gearbox associated with DGIG is studied and considered as a possible option which can limit the gearbox maintenance needs [14]. Another constraint in using a DFIG-based system is the presence of brushes and rings which allows connecting the rotor windings to the converter. These three needed options (pitch control, gearbox, and brushes and rings) are known to be a high level source of failure and maintenance is needed as shown in [2] for wind turbine context. This is why that even when this option leads to CAPEX reduction, using DFIG is less favorable to the particular context of tidal stream turbines than to the context of onshore wind turbines due to the difficulty to access to system for maintenance or repair.

Squirrel cage induction generator (SCIG) with a back-to-back full power converter that is connected to the stator windings as presented in Figure 1.25(a) can be also a possible solution to be associated with a tidal turbine. SCIGs are relatively low cost and are highly reliable. Field-oriented control of the converter allows a precise torque and speed control of the generator in a large range of rotating speed including flux weakening capabilities. However, SCIG with high efficiency and high power factor are characterized by a low number of poles (typically lower than 8). Rotor field density is also for a given cooling technology lower than in synchronous machine or in DFIG due to the fact that rotor currents are generated by induction phenomenon from the stator field. It leads to a lower tangential stress in the air gap and, as a consequence, there is a lower compactness of the machine for a given torque. Efficiency is also lower due to rotor losses. Consequently a direct-drive association of a SCIG with a MW range tidal stream turbine is difficult to consider. In fact, using a SCIG implies to use a high ratio gearbox to connect a MW turbine to the generator (typically a three-stage gearbox with a global ratio of around 100). For example, this solution has been used by MCT Company which is one of the pioneers in tidal turbine deployment for the SEAFLOW and SEAGEN projects [6,8] and also by ANDRITZ HYDRO Hammerfest for the HS1000 project and Orbital Marine Power for the O2 project (formerly scotrenewables) [7,64]. Using three-stages gearbox leads to high maintenance requirements. This is why, as for a DFIG based on solution, a SCIG have to be associated with specific systems or procedure to facilitate regular maintenance.

As an example, MCT Company has proposed lift system to extract the turbine and generator over the sea surface and to proceed to easy maintenance operations [6].

Using a synchronous generator (SG) is an interesting option for tidal stream energy harnessing. Like SCIG, SG, as shown in Figure 1.25(c), has to be associated with a full power back-to-back converter connected to stator windings which allow torque and speed control in a large range of rotational speeds, thanks to the use of field-oriented control including flux weakening operations. In a SG, rotor field can be generated using permanent magnets or windings supplied by a DC current. In terms of efficiency and torque density, using a permanent magnet synchronous generator (PMSG) is a very interesting option because rare-earth permanent magnets allow creating very high level of field density in the air gap with very low rotor losses. Thanks to this characteristic, PMSG with rare-earth magnets allows to obtain a very high value of the magnetic tangential stress (1.2) in the rotor (or stator surface). This stress can be 2 or 3 times higher than in a SCIG for example [1]. So it is possible to design a high torque low speed system with a high compactness. Using a PMSG allows a direct-drive association of the generator and the turbine. In this case, PMSG with a high number of pole pairs (more than 100) can be used. PMSG-based technologies have been chosen in several projects as for example Sabella project or Voith Hydro projects and has been studied in many academic studies as for example in [65–68]. In Figure 1.33, a picture of the PM rotor of the Sabella D10 generator under construction is shown. The high pole number of this PM rotor can be noticed in this figure. It is also possible to associate a small ratio gearbox with a PMSG in order to find a compromise between CAPEX and system reliability as explained previously in Section 1.1.2.2. As an example ATLANTIS AR 1500 uses a two-stage Gearbox associated with a PMSG [17].

Even if rare-earth permanent magnets are expensive materials and a lot of magnets are needed for the generator design, this option allows to eliminate the gearbox which

Figure 1.33   PM assembly in the rotor of SABELLA D10 generator (courtesy of SABELLA)

Table 1.2   Some of MW range tidal turbine generator and drivetrain configurations
(IG, induction generator; PMSG, permanent magnet synchronous
generator; GB, gearbox; DD: direct drive)

Company/project	Rated power	Generator and drivetrain
Andritz Hydro Hammerfest/HS100 [7]	1 MW	IG + GB (3-stage)
Atlantis-Resource Ltd /AR1500 [17]	1.5 MW	PMSG + GB (2-stage)
MCT/ Seagen S [8]	1 MW	IG + GB
Nautricity/Cormat [29]	0.5 MW	PMSG DD (contra-rotating)
Scotrenewables/SR2000 [69]	2 × 1 MW	IG + GB
SABELLA/D10 [36]	1 MW	PMSG DD

is expected to be one of the main maintenance needs and failure source. This is why even when the CAPEX of the generator is high (due to PM) and when integrating the generator in a nacelle without affecting the hydrodynamic efficiency is challenging, using a PMSG based on system can be a very relevant technical solution in the tidal stream energy harnessing context. Considering the previously described solutions and the specific constraints and design goals related to tidal stream energy harnessing, the main MW size tidal projects will use massively PMSG and IG as shown in Table 1.2 which describes a few MW size project generator options for horizontal axis MW range tidal turbine.

IG with gearbox is an interesting solution to try to find an optimal compromise between reliability and cost and using PMSG (direct drive or with limited ratio gearbox) is a good solution to improve the whole system reliability.

1.2.3   Generator and drive flux weakening specifications for fixed pitch turbine

As explained in previous paragraphs, it is possible to use a MPPT combined with power limitation harnessing strategy for fixed pitch turbines. This solution can be used with over or underspeed power limitation strategy. It is therefore obvious, if we consider the torque-speed characteristics for these two cases (Figures 1.32 and 1.31), that the design of a generator-drive combination will allow the maximum steady-state torque ($T_{\max}$) to be controlled for the maximum rotational speed ($\Omega_{\max}$), which will result in a considerable oversizing of the converter (the rated power of the converter will be in this case more than twice the rated power of the turbine). A possible solution to limit the converter size is to design a generator and drive association with relevant flux weakening capabilities. Basically using a flux weakening strategy will allow to control torque lower or equal than rated maximal torque ($T_{g\max}$) until a base rotating speed (Ω_b) and to control a power lower than approximately the maximal constant rated power ($P_{g\max}$) which is equal to the power reached at the base point ($T_{g\max}$, Ω_b) in a limited speed range when the rotating speed is greater than the base speed. This torque VS speed capability is described by (1.8) and (1.9). Which means that all the torque and speed values lower than the torque limits defined by these two equations

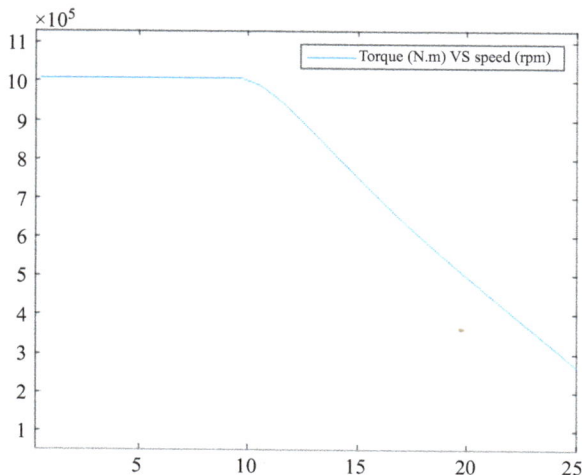

Figure 1.34   Example of torque VS speed capability of an AC generator with flux weakening control

are possible operating points for the system. This kind of strategy is for example massively used for electrical and hybrid vehicle motors. This kind of strategy can be used in both SG and IG but requires specific design option:

$$\Omega < \Omega_b : T < T_{gmax} \tag{1.8}$$

$$\Omega_{gmax} > \Omega > \Omega_b : T \lessgtr \frac{P_{gmax}}{\Omega} \tag{1.9}$$

where T_{gmax} and P_{gmax} are the rated torque and power of the generator and drive association, respectively. Ω_b and Ω_{gmax} are the base speed and the maximal speed of the generator and drive, respectively.

In Figure 1.34, a typical example of flux weakening characteristics of a PMSG or IG is presented. In this figure, the base point corresponds to approximately 1 MW for 10 rpm.

Flux weakening strategy can then be used to optimize the sizing of both generator and drive in order to operate the system for all the operating points corresponding to overspeed or underspeed power limitation strategy [55,62]. Designing the generator and drive system for a base speed maximal operating point a little over than the turbine maximal torque operating point (point 2 in Figure 1.31 or point 1 in Figure 1.32) can be an interesting option to design an optimized size converter and generator solution as shown in Figure 1.35. Generator and drive design has then to be optimized to be able to operate in all the operating points. As an example, in [55], a PMSG DD generator and its back-to-back converter are optimized in terms of active part costs to be able to fit with a given turbine control strategy including MPPT and overspeed power limitation harnessing strategy.

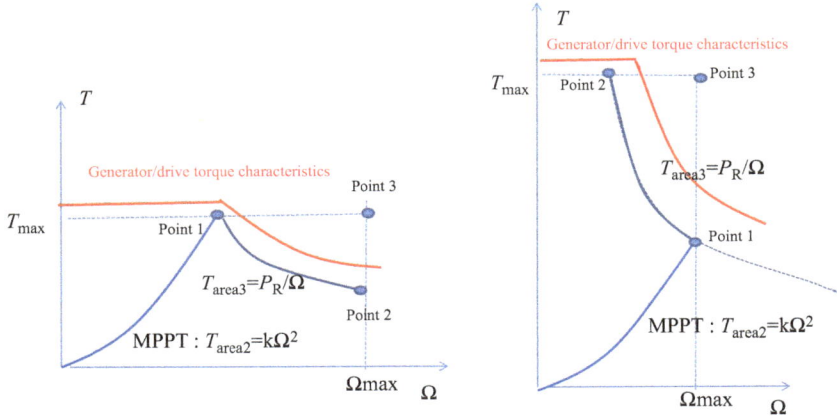

Figure 1.35   Possible flux weakening characteristics and turbine operating points in the torque VS speed frame for overspeed (right) and underspeed (left) power limitation strategies with a fixed pitch tidal turbine

1.2.4   Focus on direct-drive PM generators dedicated to TST

On the one hand, IGs used for TSTs correspond to relatively classical induction machines because they correspond to relatively high speed low torque machines thanks to the use of high ratio (three-stage) gearboxes. On the other hand, direct-drive PMSGs correspond to very unconventional electrical machine specifications: generators have to be operated at very high torque (MN m range) and low-speed and have to be integrated with the turbine blades without affecting significantly the turbine hydrodynamic performance. They must also be as reliable and as cheap as possible. This is why original solutions have been proposed in the literature and in some projects to face these challenges. Some of them will be presented in the next part of this paragraph.

1.2.4.1   Rim-driven PMSG

It is obvious considering (1.2) and Figure 1.3 that DD PMSG tidal generator that is in the MN.m range in terms of rated torque will lead to very huge rotor external volume and consequently huge external volume. As an example, a 1 MW, 12 rpm will have to consider (1.2) and a typical PMSG tangential stress of about 35 kPa will have an external rotor volume of about 11 m^3. However, as the number of poles will be very high (typically more than 100 poles), stator and rotor core height will be very small (typically a few centimeters) for rotor diameters greater than 2 m. That means that the interior of the rotor is free of electrical parts. As an example, the sketch of a 300 kW/15 rpm PMSG with an external diameter of 2 m is presented in Figure 1.36. This PMSG has been designed for a 11 m diameter tidal turbine in [70].

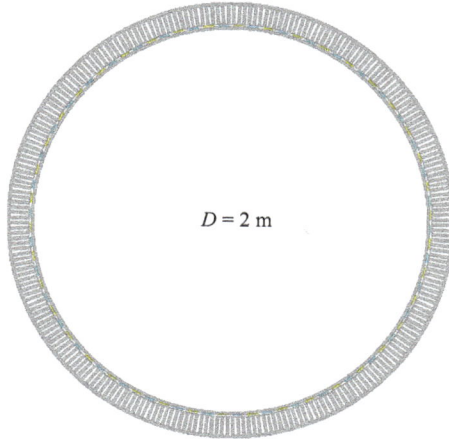

Figure 1.36   Example of a PMSG designed for tidal turbine (300 kW/15 rpm) to be included in a nacelle in the turbine axis [70]

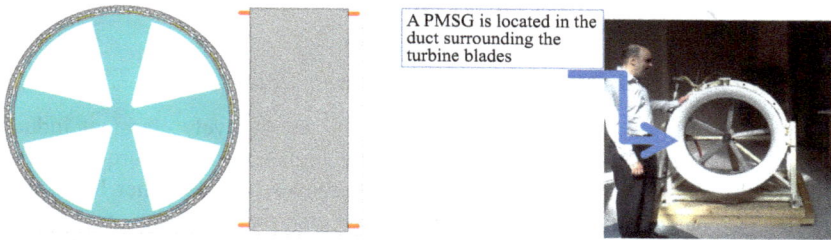

Figure 1.37   Principle of a Rim-driven system and example of a laboratory prototype of RIM-driven turbine (courtesy from IRENAV)

By considering this assumption, it can be shown that increasing diameter will lead to active mass volume reduction [67]. So a possible solution is to include the PMSG in a duct surrounding the blade [65]. In this configuration, the rotor active parts (iron core and magnets) are directly fixed to the external part of the blades as shown in Figure 1.37. In some studied configuration, the air gap can be filled with seawater.

This solution that has also been employed for naval propulsion systems [71] allows theoretically optimizing active mass and cost and is also expected to improve the hydrodynamic efficiency. This technical option has been chosen by DCNS-OPENHYDRO project which have been stopped in 2018 [60] and by Torcado Company between 1999 and 2005 [72]. However, maintaining the rotor in the stator is challenging and implies to use specific peripheral bearing. The presence of seawater in the airgap also leads to strong sealing constraints. Very small generator active length leads to also relatively poor generator efficiency due to huge relative losses in end-windings [73]. So this option is not fully mature for MW range tidal system and several studies are conducted to try to optimize the behavior of RIM-driven tidal generator. As examples, the use of an axial flux PMSG is proposed in [27] and the use of segmented stator PMSG for Rim-driven tidal turbines is presented in [73].

1.2.4.2 Axial flux generators

Axial flux PM synchronous machines are an interesting possibility to increase both the generator compactness and reliability. This is why an axial flux PMSG can be relevant in the context of tidal stream energy harnessing. Axial flux generators are made with several discoidal parts facing each other. Some of them (stator disks) are carrying windings and some of them are carrying permanent magnets (rotor disks). The flux lines are then oriented in axial direction in the air gaps. Figure 1.38 presents some topology of axial flux permanent magnet machine configurations. The particular configurations of axial flux machine allow increasing the surface of the air gaps which is the area of interaction of stator and rotor field and offers bigger thermal exchange surfaces for machine cooling. This is why these machines are expected to be more compact than more classical radial flux machines. As an example, the study

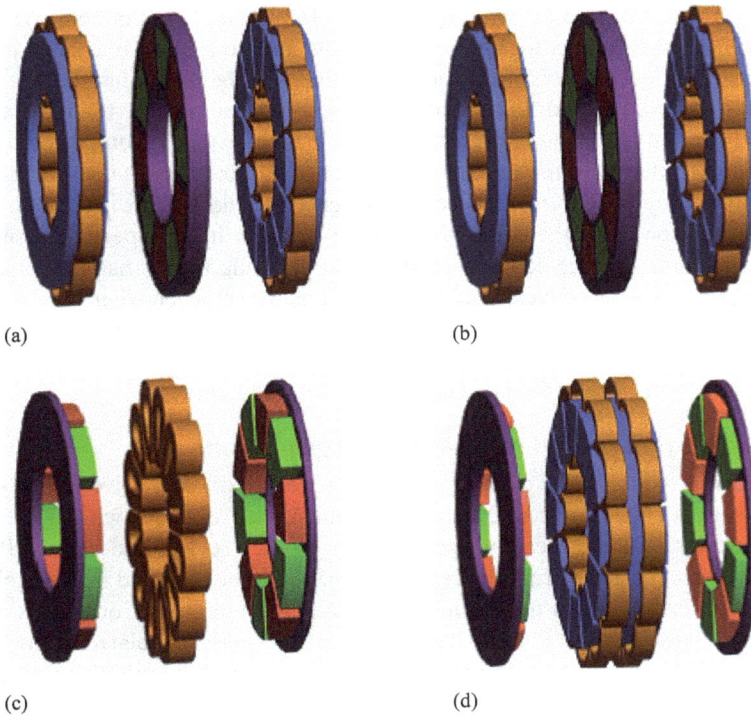

(a)

(b)

(c)

(d)

Figure 1.38 Possible topologies of AFPM machines with three disks: (a) and (b) 2 stators/1 rotor and (c) and (d) 2 stators/1 rotor

presented in [27] shows that using a double stator single rotor structure allows to reduce significantly mass and volume of active parts for RIM-driven configurations. The multidisc configurations also allow increasing reliability offering degrees of freedom in terms of fault tolerance configuration [67]. As an example, a double stator axial PM machine can be operated at half rated power even when one of the stator windings is out of order, if each of the stators is associated with an independent converter. If several rotors are used, the machine design can be configured to operate with two rotors in contra-rotating sense which allows increasing the hydrodynamic efficiency of the system. For example, this solution has been used for the CoRMat contra-rotating turbine that uses a contra-rotating double rotor axial field generator [29]. However, manufacturing aspects can limit the development of large axial flux PM structures due to the difficulties to build laminated magnetic core in axial direction and to maintain mechanically the air gaps and the disk position. This is why the radial flux generators are used in the main part of MW size turbine based on direct-drive PMSG.

1.2.4.3   Unconventional PMS generators based on saliency effects

Some unconventional PMSG configurations are based on PMSG which use saliency effects. As an example double saliency permanent magnet generator (DSPMG) is proposed by some academic studies in order to increase PMSG compactness. In this kind of machine, PM and AC windings are both located in the stator core. Windings are concentrated windings located in stator plots and both rotor and stator present teethes which allow a saliency variation when the machine is running. This saliency variation creates a no load flux variation in the windings in order to create a back EMF. This kind of machine is expected to have a high level and compactness and a high reliability, thanks to the particular structure of the rotor which does not contain field source. They have been proposed and designed for tidal turbine in [23,74]. Some studies have shown that the use of this kind of machine in low-speed generators is expected to minimize the active part mass and particularly the magnet volume in comparison with more conventional PMSG [75]. However, even when they are very promising for tidal stream turbine generator applications, these machine technologies are not yet fully mature and some research works are conducted in order to solve issues in terms of control, power factor, and manufacturing [74,76,77].

1.2.4.4   Multiphase and multi-windings generators

As increasing reliability of tidal stream turbine generators and drive is one of the major issues for these systems, the use of multi-phase or multi-windings PMSG can be relevant options. It can be used with any of the previously described PMSG structures. A multiphase or multi-windings PMSG can be operated in fault tolerant mode even when a part of the converter or windings is in fault. Constraints on the switches are also reduced thanks to a derating of the switches related to the power splitting. A first option is to design and use regularly shifted multiphase machines (with a phase number greater than 3). As examples, some studies have investigated the use of 5-phase PMSG for tidal turbine in order to be operated in faulted conditions as in [78,79]. This study has been extended to the use of flux weakening strategy in faulted operations in [80–82]. Figure 1.39 shows the principle scheme of a 5-phase

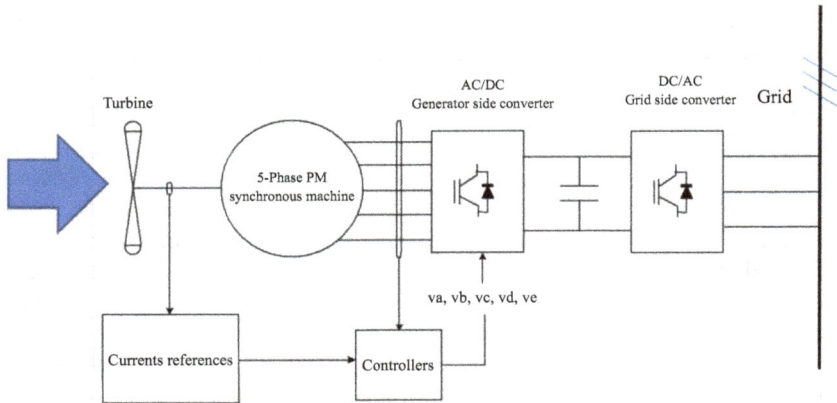

Figure 1.39   5-Phase PMSG and drive scheme

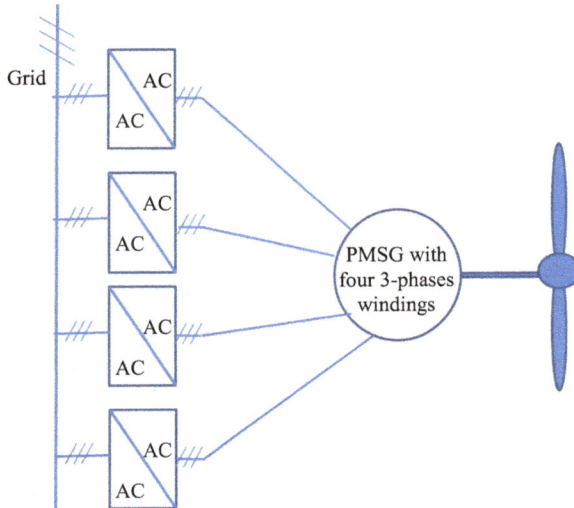

Figure 1.40   Four-stars winding PMSG and drive scheme

PMSG and drive. Another interesting characteristic of multi-phase machine is that harmonics injection in the current can allow to increase significantly torque density and so to increase the compactness of the machine, thanks to the supplementary degree of freedom offered by phase number greater than 3. In [83], it has been demonstrated that in a 5-phase PMS machine, for the same joule losses and design constraints, an increase of the torque of 15% can be expected using third harmonic current injection, in comparison with an equivalent classical 3-phase machine using sinusoidal current control.

However, using a regularly shifted multiphase system leads to use an unconventional converter and current control strategy. Another solution is to use multi 3-phase windings PMSG (multi-star windings). In this configuration, the PMSG stator windings are made with several independent 3-phase windings which can be associated each other to an independent converter as shown in Figure 1.40 which presents a 4×3-phases PMSG. It allows using mature and on the shield 3-phases converters and control strategies. In this configuration, if a default is detected in one of the winding or associated converter. This winding can be disconnected and the system can then be operated with reduced power with the remaining healthy windings. This solution, which is classically used for high power ship propulsion, has been for example studied in terms of reliability for a realistic tidal turbine mission profile [84]. This study shows that a multi-star solution can be a very relevant solution in terms of compromise between reliability and performance in the whole lifetime of a tidal turbine.

1.3   Summary and conclusion

This chapter has presented several design options for generators and drivetrains, which can be used for tidal stream generation to reach the challenging requirements of tidal energy generation in terms of cost, reliability, efficiency, and system integration: geared and direct-drive options (including rim-driven option) generators with several possible technological options (induction generators, synchronous generators, doubly fed induction generators, unconventional generators). The use of possible technological choices for gearboxes was also discussed, including conventional and specific magnetic gearboxes, which are a relevant way to improve powertrains in this context. Consequences of generators and drivetrains choices when using a specific optimal harnessing strategy (MPPT with power leveling) associated with several turbine technological options (horizontal or vertical axis turbines, fixed- or variable-pitch turbines) have also been discussed. This chapter will undoubtedly serve as a guide for generator and drivetrain choice which can be useful for prospective tidal turbine developers considering the main constraints related to the marine energy harnessing, including cost (i.e., OPEX and CAPEX), reliability, and efficiency.

References

[1]   Jokinen T, Hrabovcova V, and Pyrhonen J. *Design of Rotating Electrical Machines*. John Wiley & Sons, New York, NY; 2013.

[2]   Ribrant J and Bertling L. Survey of failures in wind power systems with focus on Swedish wind power plants during 1997–2005. In: *2007 IEEE Power Engineering Society General Meeting*. IEEE; 2007. p. 1–8.

[3]   Touimi K, Benbouzid M, and Tavner P. Tidal stream turbines: with or without a gearbox? *Ocean Engineering*. 2018;170:74–88.

[4]   Faris Elasha DM and Teixeira JA. Condition monitoring philosophy for tidal turbines. *International Journal of Performability Engineering*. 2014;10(5):521–534.

[5] SEAFLOW World's First Pilot Project for the Exploitation of Marine Currents at a Commercial Scale. European Commission; 2005.

[6] Fraenkel P. Practical tidal turbine design considerations: a review of technical alternatives and key design decisions leading to the development of the SeaGen 1.2 MW tidal turbine. In: *Ocean Power Fluid Machinery Seminar*, vol. 19; 2010. p. 1–19.

[7] Hammerfest AH. Andritz Hydro Hammerfest Turbine Brochure. https://www.andritz.com/resource/blob/61614/cf15d27bc23fd59db125229506ec87c7/hy-hammerfest–1–data.pdf [cited 2022 June 28].

[8] Ltd MCT. SeaGen–S Brochure. https://atlantisresourcesltd.com/wp/wp-content/uploads/2016/08/SeaGen-Brochure.pdf [cited 2022 June 28].

[9] Touimi K, Benbouzid M, and Chen Z. Optimal design of a multibrid permanent magnet generator for a tidal stream turbine. *Energies*. 2020; 13(2):487.

[10] Multibrid M5000. https://en.wind-turbine-models.com/turbines/22-multibrid-m5000 [cited 2022 June 28].

[11] Marsh G. Turbine producers step into ten league boots. *Renewable Energy Focus*. 2010;11(4):46–51.

[12] Lin Y, Tu L, Liu H, *et al*. Fault analysis of wind turbines in China. *Renewable and Sustainable Energy Reviews*. 2016;55:482–490.

[13] Carroll J, McDonald A, Feuchtwang J, *et al*. Drivetrain availability in off-shore wind turbines. In: *European Wind Energy Association 2014 Annual Conference*; 2014.

[14] Polinder H, Van der Pijl FF, De Vilder GJ, *et al*. Comparison of direct-drive and geared generator concepts for wind turbines. *IEEE Transactions on Energy Conversion*. 2006;21(3):725–733.

[15] Liu H and Bahaj AS. Status of marine current energy conversion in China. *International Marine Energy Journal*. 2021;4(1):11–23.

[16] Ma S. *Study on Energy Conversion Efficiency and Power Control of Horizontal Axis Tidal Current Energy Conversion Systems*. Zhejiang University; 2011.

[17] Energy SA. AR1500 Tidal Turbine Brochure. https://simecatlantis.com/wp/wp-content/uploads/2016/08/AR1500-Brochure-Final-1.pdf [cited 2022 June 28].

[18] Matveev A. Novel PM generators for large wind turbines. In: *Wind Power R&D Seminar—Deep Sea Offshore Wind Power*; 2011.

[19] Keysan O. Future electrical generator technologies for offshore wind turbines. *Engineering & Technology (E&T) Reference*. 2015;1(1):1–14.

[20] Keysan O, McDonald A, Mueller M, *et al*. C-GEN, a lightweight direct drive generator for marine energy converters. In: *IET Power Electronics Machines & Drives Conference*; 2010.

[21] Keysan O, McDonald AS, and Mueller M. A direct drive permanent magnet generator design for a tidal current turbine (SeaGen). In: *2011 IEEE International Electric Machines & Drives Conference (IEMDC)*. IEEE; 2011. p. 224–229.

[22] Jin J, Charpentier JF, and Tang T. Preliminary design of a TORUS type axial flux generator for direct-driven tidal current turbine. In: *2014 First International Conference on Green Energy ICGE 2014*. IEEE; 2014. p. 20–25.

[23] Harkati N, Moreau L, Zaim M, *et al.* Low speed doubly salient permanent magnet generator with passive rotor for a tidal current turbine. In: *2013 International Conference on Renewable Energy Research and Applications (ICRERA)*. IEEE; 2013. p. 528–533.

[24] Funieru B and Binder A. Design of a PM direct drive synchronous generator used in a tidal stream turbine. In: *2013 International Conference on Clean Electrical Power (ICCEP)*. IEEE; 2013. p. 197–202.

[25] Chen H, At-Ahmed N, Machmoum M, *et al.* Modeling and vector control of marine current energy conversion system based on doubly salient permanent magnet generator. *IEEE Transactions on Sustainable Energy*. 2015;7(1): 409–418.

[26] Baker NJ, Cawthorne S, Hodge E, *et al.* *3D Modelling of the Generator for OpenHydro's Tidal Energy System*. IET Press, UK; 2014.

[27] Djebarri S, Charpentier JF, Scuiller F, *et al.* Design and performance analysis of double stator axial flux PM generator for rim driven marine current turbines. *IEEE Journal of Oceanic Engineering*. 2015;41(1):50–66.

[28] Clarke JA, Connor G, Grant A, *et al.* Contra-rotating marine current turbines: single point tethered floating system-stability and performance. In: *8th European Wave and Tidal Energy Conference*, EWTEC 2009; 2009.

[29] Porter K, Ordonez-Sanchez S, Johnstone C, *et al.* Integration of a direct drive contra-rotating generator with point absorber wave energy converters. In: *12th European Wave and Tidal Energy Conference*; 2017.

[30] Clarke JA, Connor G, Grant A, *et al.* Design and testing of a contra-rotating tidal current turbine. *Proceedings of the Institution of Mechanical Engineers, Part A: Journal of Power and Energy*. 2007;221(2):171–179.

[31] Nautricity Ltd. and Fundy Tidal Inc. have Signed an MoU to Develop a 500 kW Tidal Project in Nova Scotia, Canada. https://www.offshorewind.biz/2014/06/25/nautricity-fundy-tidal-to-develop-tidal-energy-project-in-nova-scotia-canada/ [cited 2022 October 30].

[32] Open Hydro 250 kW Prototype. https://www.emec.org.uk/about-us/our-tidal-clients/open-hydro/ [cited 2022 June 28].

[33] Race Rocks Demonstration Project. https://web.archive.org/web/20080705173 021/http://www.cleancurrent.com/technology/rrproject.htm [cited 2022 June 28].

[34] Press TC. *Cape Sharp Tidal Turbine in Bay of Fundy Now Being Monitored Remotely*. The Canadian Press; 2018. https://www.cbc.ca/news/canada/nova-scotia/cape-sharp-tidal-turbine-remote-monitoring-environment-1.4814069 [cited 2022 June 28].

[35] Voith's 1 MW Horizontal Axis TST. https://www.emec.org.uk/about-us/our-tidal-clients/voith-hydro/ [cited 2022 June 28].

[36] Paboeuf S, Yen Kai Sun P, Macadré LM, *et al.* Power performance assessment of the tidal turbine Sabella D10 following IEC62600-200. In: *International*

Conference on Offshore Mechanics and Arctic Engineering, vol. 49972. American Society of Mechanical Engineers; 2016. p. V006T09A007.

[37] Gerber S and Wang R. Evaluation of a prototype magnetic gear. In: *2013 IEEE International Conference on Industrial Technology (ICIT)*. IEEE; 2013. p. 319–324.

[38] Rasmussen PO, Andersen TO, Jorgensen FT, *et al.* Development of a high-performance magnetic gear. *IEEE Transactions on Industry Applications*. 2005;41(3):764–770.

[39] Shah L, Cruden A, and Williams BW. A magnetic gear box for application with a contra-rotating tidal turbine. In: *2007 7th International Conference on Power Electronics and Drive Systems*. IEEE; 2007. p. 989–993.

[40] Atallah K and Howe D. A novel high-performance magnetic gear. *IEEE Transactions on Magnetics*. 2001;37(4):2844–2846.

[41] Bronn L, Wang R, and Kamper M. Development of a shutter type magnetic gear. In: *Proceedings of the 19th Southern African Universities Power Engineering Conference*; 2010. p. 78–82.

[42] Frank NW and Toliyat HA. Analysis of the concentric planetary magnetic gear with strengthened stator and interior permanent magnet inner rotor. *IEEE Transactions on Industry Applications*. 2011;47(4):1652–1660.

[43] Fukuoka M, Nakamura K, and Ichinokura O. Experimental tests of surface permanent magnet magnetic gear. In: *2012 15th International Conference on Electrical Machines and Systems (ICEMS)*. IEEE; 2012. p. 1–6.

[44] Magnomatics. *Tidal Current Generator*. https://www.magnomatics.com/renewable-ocean-energy [cited 2022 June 28].

[45] Dragan RS, Barrett R, Calverley S, *et al.* Pseudo-direct-drive electrical machine for a floating marine turbine. *IEEE Transactions on Magnetics*. 2021;58(2):1–5.

[46] Yin X, Fang Y, and Pfister PD. High-torque-density pseudo-direct-drive permanent-magnet machine with less magnet. *IET Electric Power Applications*. 2017;12(1):37–44.

[47] Atallah K, Rens J, Mezani S, *et al.* A novel "pseudo" direct-drive brushless permanent magnet machine. *IEEE Transactions on Magnetics*. 2008;44(11): 4349–4352.

[48] Penzkofer A and Atallah K. Scaling of pseudo direct drives for wind turbine application. *IEEE Transactions on Magnetics*. 2016;52(7):1–5.

[49] Magnomatics. *Compact High Efficiency Generator for Wind Turbines*. Available from: https://www.magnomatics.com/post/wind-turbine-generator [cited 2022 September 25].

[50] Jian L, Chau K, and Jiang J. A magnetic-geared outer-rotor permanent-magnet brushless machine for wind power generation. *IEEE Transactions on Industry Applications*. 2009;45(3):954–962.

[51] Johnson M, Gardner MC, and Toliyat HA. Design and analysis of an axial flux magnetically geared generator. *IEEE Transactions on Industry Applications*. 2016;53(1):97–105.

[52]   Benelghali S. *On Multiphysics Modeling and Control of Marine Current Turbine Systems*. Université de Bretagne occidentale-Brest; 2009.

[53]   El Tawil T, Charpentier JF, and Benbouzid M. Tidal energy site characterization for marine turbine optimal installation: case of the Ouessant Island in France. *International Journal of Marine Energy*. 2017;18:57–64.

[54]   El Tawil T, Guillou N, Charpentier JF, *et al*. On tidal current velocity vector time series prediction: a comparative study for a French high tidal energy potential site. *Journal of Marine Science and Engineering*. 2019;7(2):46.

[55]   Djebarri S, Charpentier JF, Sciuller F, *et al*. Design methodology of permanent magnet generators for fixed-pitch tidal turbines with overspeed power limitation strategy. *Journal of Ocean Engineering and Science*. 2020;5(1):73–83.

[56]   Zhou Z. *Modeling and Power Control of a Marine Current Turbine System with Energy Storage Devices*. Université de Bretagne occidentale-Brest; 2014.

[57]   Benelghali S, Benbouzid MEH, and Charpentier JF. Modelling and control of a marine current turbine-driven doubly fed induction generator. *IET Renewable Power Generation*. 2010;4(1):1–11.

[58]   Bahaj A, Molland A, Chaplin J, *et al*. Power and thrust measurements of marine current turbines under various hydrodynamic flow conditions in a cavitation tunnel and a towing tank. *Renewable Energy*. 2007;32(3):407–426.

[59]   Allsop S, Peyrard C, Thies PR, *et al*. Hydrodynamic analysis of a ducted, open centre tidal stream turbine using blade element momentum theory. *Ocean Engineering*. 2017;141:531–542.

[60]   Zhou Z, Benbouzid M, Charpentier JF, *et al*. Developments in large marine current turbine technologies—a review. *Renewable and Sustainable Energy Reviews*. 2017;71:852–858.

[61]   Arnold M, Biskup F, and Cheng PW. Load reduction potential of variable speed control approaches for fixed pitch tidal current turbines. *International Journal of Marine Energy*. 2016;15:175–190.

[62]   Zhou Z, Sciuller F, Charpentier JF, *et al*. Power control of a nonpitch-able PMSG-based marine current turbine at overrated current speed with flux-weakening strategy. *IEEE Journal of Oceanic Engineering*. 2014;40(3): 536–545.

[63]   Djebarri S, Charpentier JF, Sciuller F, *et al*. Influence of fixed-pitch tidal turbine hydrodynamic characteristic on the generator design. In: *11th European Wave and Tidal Energy Conference*. EWTEC; 2015. p. 8A2–2.

[64]   plc OMPO. *Project Information Summary*. https://marine.gov.scot/sites/default/files/project_information_summary_4.pdf [cited 2022 June 28].

[65]   Drouen L, Charpentier JF, Semail E, *et al*. Study of an innovative electrical machine fitted to marine current turbines. In: *OCEANS 2007-Europe*. IEEE; 2007. p. 1–6.

[66]   Wani FM. *Improving the Reliability of Tidal Turbine Generator Systems*. Delft University of Technology; 2021.

[67]   Djebarri S, Charpentier JF, Sciuller F, *et al*. Comparison of direct-drive PM generators for tidal turbines. In: *2014 International Power Electronics and Application Conference and Exposition*. IEEE; 2014. p. 474–479.

[68]   Benelghali S, Benbouzid MEH, and Charpentier JF. Generator systems for marine current turbine applications: a comparative study. *IEEE Journal of Oceanic Engineering*. 2012;37(3):554–563.

[69]   Ltd STP. SR2000 Brochure. https://cdn.ymaws.com/www.renewableuk.com/resource/resmgr/docs/health_&_safety/scotrenewables_safe_access_f.pdf [cited 2022 September 20].

[70]   Djebarri S. *Contribution à la modélisation et à la conception optimale de génératrices à aimants permanents pour hydroliennes*. Université de Bretagne occidentale-Brest; 2015.

[71]   Yan X, Liang X, Ouyang W, *et al*. A review of progress and applications of ship shaft-less rim-driven thrusters. *Ocean Engineering*. 2017;144:142–156.

[72]   History of Tocardo. https://www.tocardo.com/about/history/ [cited 2022 June 28].

[73]   Fleurot E, Charpentier JF, and Scuiller F. Electromagnetic study of segmented permanent magnet synchronous machines for Rim-driven applications. In: *2019 19th International Symposium on Electromagnetic Fields in Mechatronics, Electrical and Electronic Engineering (ISEF)*. IEEE; 2019. p. 1–2.

[74]   Harkati N, Moreau L, Zaïm M, *et al*. Optimized design of doubly salient permanent magnet generator taking into account converter constraints. In: *2016 International Conference on Electrical Sciences and Technologies in Maghreb (CISTEM)*. IEEE; 2016. p. 1–8.

[75]   Guerroudj C, Charpentier JF, Saou R, *et al*. Coil number impact on performance of 4-phase low speed toothed doubly salient permanent magnet motors. *Machines*. 2021;9(7):137.

[76]   Chen H, Tang S, Han J, *et al*. High-order sliding mode control of a doubly salient permanent magnet machine driving marine current turbine. *Journal of Ocean Engineering and Science*. 2021;6(1):12–20.

[77]   Chen H, Tang T, Han J, *et al*. One special current waveform of toothed pole doubly salient permanent magnet machine for marine current energy conversion system. *Electrical Engineering*. 2020;102(1):371–386.

[78]   Mekri F, Elghali SB, and Benbouzid MEH. Fault-tolerant control performance comparison of three-and five-phase PMSG for marine current turbine applications. *IEEE Transactions on Sustainable Energy*. 2012;4(2):425–433.

[79]   Pham HT, Bourgeot JM, and Benbouzid M. Fault-tolerant model predictive control of 5-phase PMSG under an open-circuit phase fault condition for marine current applications. In: *IECON 2016—42nd Annual Conference of the IEEE Industrial Electronics Society*. IEEE; 2016. p. 5760–5765.

[80]   Fall O, Nguyen NK, Charpentier JF, *et al*. Variable speed control of a 5-phase permanent magnet synchronous generator including voltage and current limits in healthy and open-circuited modes. *Electric Power Systems Research*. 2016;140:507–516.

[81]   Fall O, Charpentier JF, Nguyen NK, *et al*. Performances comparison of different concentrated-winding configurations for 5-phase PMSG in normal and faulty modes in flux weakening operation for fixed pitch tidal turbines. In: *2016*

XXII International Conference on Electrical Machines (ICEM). IEEE; 2016. p. 2789–2795.

[82]   Fall O, Charpentier JF, Nguyen NK, *et al*. Maximum torque per ampere control strategy of a 5-phase PM generator in healthy and faulty modes for tidal marine turbine application. In: *2014 International Power Electronics and Application Conference and Exposition*. IEEE; 2014. p. 468–473.

[83]   Scuiller F, Semail E, Charpentier JF, *et al*. Multi-criteria-based design approach of multi-phase permanent magnet low-speed synchronous machines. *IET Electric Power Applications*. 2009;3(2):102–110.

[84]   Olmi C, Scuiller F, and Charpentier JF. Impact of a multi-star winding on the reliability of a permanent magnet generator for marine current turbine. *International Journal of Marine Energy*. 2017;19:319–331.

Chapter 2
Tidal stream turbine control
Hafiz Ahmed[1], Hao Chen[2] and Mohamed Benbouzid[3]

Tidal stream turbines (TST) operate in a harsh marine condition. TSTs interact with the tide to produce electricity, where the tide rotates the turbine and the turbine in return rotates the electrical generator for electricity production as shown in Figure 2.1. In order to maximise the energy production from the tide and also to make the system robust to various disturbances and parameter uncertainties, control systems play a great role. Development of a control system for tidal stream turbines is the main focus of this chapter. Due to the submerged and/or semi-submerged operation, TSTs face several control changes. First, an uncertain marine environment can introduce large amplitude external disturbances, which makes the turbine speed control very challenging. Grid-interactive power electronic converters are used to export the TST-generated electrical power to the grid. Control of the converter(s) requires accurate information of the system parameters. However, marine environments can cause significant degradation of the system, resulting in rapid change of system parameters. This makes controlling the converter particularly difficult as the conventional controller often requires accurate value of the system parameters. Finally, TSTs are often located in remote locations where the grid may be weak, i.e., the grid may have low inertia. Integrating TSTs into a weak grid adds additional control challenges from the grid-connection view point. This chapter provides an overview of TSTs control methods that address these challenges.

This chapter begins with a short overview of control methods in Section 2.1, where one linear and one non-linear control method are described in detail. Mathematical models of the individual components of the tidal stream turbines are presented in Section 2.2, where tidal resources, tidal turbines, and electrical generators are modelled. Development of control methods for different types of electrical generator-based TSTs are given in Section 2.3. This section also presents the validation of the controllers for different types of generator-based tidal stream turbine systems. Finally, some concluding remarks are given in Section 2.4.

[1]Bangor University, Nuclear Futures Institute, UK
[2]Shanghai Maritime University, Logistics Engineering College, China
[3]University of Brest, CNRS, Institut de Recherche Dupuy de Lôme, France

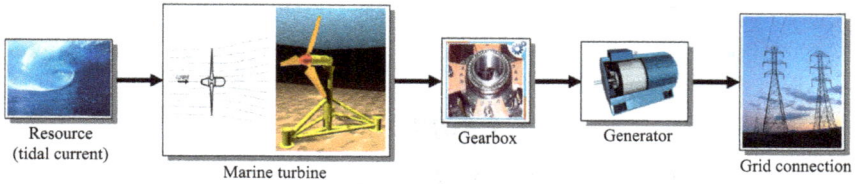

Figure 2.1   *The global scheme of TST [1]*

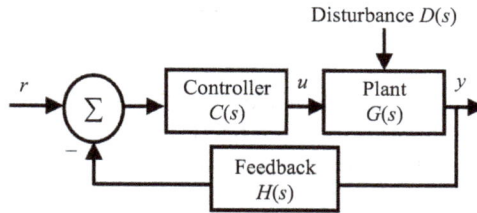

Figure 2.2   *Overview of feedback control system*

2.1   Overview of control system

Feedback control is vital to ensure appropriate operation of the tidal energy system. A general overview of a control system with negative feedback is given in Figure 2.2. The controller, which is represented by transfer function $C(s)$, compares the reference signal (r) with the output signal (y) obtained through the feedback element $H(s)$ and generates the control output u which is a function of the error $e = r - y$. The output of the controller is passed to the plant, which is represented by $G(s)$ and is the tidal turbine in our case. This turbine is subjected to various disturbances and they are represented by the transfer function $D(s)$. The purpose of controller $C(s)$ is to ensure that the output y follows the reference r despite the presence of external disturbances and/or parameter mismatches. This has motivated the development of numerous control methods for tidal stream turbines. Some popular control methods are proportional-integral-derivative (PID) [2], fuzzy logic control (FLC) [3], state-feedback control through linear quadratic regulator (LQR) [4], active disturbance rejection control (ADRC) [5], sliding mode control (SMC) [6], etc. to name a few.

The developed controllers are broadly classified into two categories, namely linear methods and nonlinear methods. PID, LQR, ADRC, etc. are some of the widely known linear methods. Conventionally, PID controller is the most popular choice in the literature. This method can be tuned in a model-free manner. However, the performance deteriorates in the presence of disturbance. Moreover, the tuning is often done by assuming a worst-case disturbance scenario. This can make tuning conservative, resulting in sub-optimal control energy utilization. Similar problems can be attributed to LQR. However, ADRC estimates the disturbance in real-time and, therefore, the designed controller is non-conservative. It is to be noted here that ADRC comes with the additional complexity of designing an estimator. Moreover, multiple control-loops are needed for the complete control of TST. Analyzing the interaction between these controllers in the presence of disturbance estimator is not straightforward. Initially linear ADRC was proposed; however, later on nonlinear ADRC [7] was also reported.

Considering the particular control challenges faced by TSTs, several nonlinear controllers are applied to TST system. In general, nonlinear methods often provide enhanced robustness compared to their linear counterpart. As highlighted previously, PID controllers suffer in the presence of disturbance. To overcome this issue, fuzzy logic controller is added together with the PID controller in [3]. In addition to the additional computational complexity, this method suffers from lack of systematic gain tuning process. Fuzzy logic operates by fuzzification of the input variables and the control law calculation requires de-fuzzification. There are no systematic rules for this purpose. This can make the tuning very plant and operation condition specific.

In the control literature, SMC [6,8,9] is often considered as a very suitable tool to provide robust stabilization and/or tracking of uncertain nonlinear systems. Since tidal turbines often experience various nonlinear disturbances and suffer from parametric uncertainties, SMC method has attracted wider attention for this kind of systems. Numerous results are already available in the literature on the application of SMC technique for TSTs. Although initially, conventional first-order SMC gained popularity; however, in recent times, high-order SMC methods started to receive more attention. Unlike the first-order counterpart, high-order SMC methods can often reduce chattering or high-frequency oscillation, which is one of the major limitation of SMC. In addition, high-order SMCs can often provide finite- and or fixed-time convergence, which is highly desirable for TSTs.

Based on the literature review presented here, two methods are selected for further discussion. They are PID and SMC. These methods will be reviewed in the following to give the readers a better picture about these two popular methods. To further contextualize the importance of control system, the general control block diagram of a TST is given in Figure 2.3.

2.1.1 PID control

PID and its various variants e.g. proportional-integral (PI), proportional-derivative (PD), etc. are undoubtedly the most popular control methods across a wide range of technical systems including tidal stream turbine. More than 90% of the controllers that are used in industry are a PID-type controller. Following Figure 2.2, the output of a PID controller is given by:

$$u(t) = K_p e(t) + K_i \int e(t)dt + K_d \frac{de(t)}{dt} \tag{2.1}$$

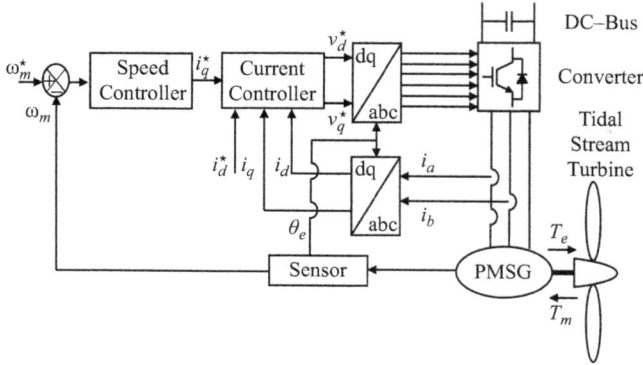

Figure 2.3   Control system overview of a TST [10]

where K_p, K_i, and K_d are the tuning gains of proportional, integral, and derivative terms, respectively. In Laplace domain, the transfer function of the PID controller can be written as:

$$C(s) = \frac{U(s)}{E(s)} = K_p + \frac{K_i}{s} + K_d s \tag{2.2}$$

where s is the Laplace operator. The controller equation in terms of time constants can be written as:

$$u(t) = K_p \left(e(t) + \frac{1}{T_i} \int e(t)dt + T_d \frac{de(t)}{dt} \right) \tag{2.3}$$

where T_i and T_d are integral and derivative time constants. An overview of feedback control system with PID controller is given in Figure 2.4. PID controller is a three-term controller, where each term is responsible collectively to improve the overall system behaviour and enhance the stability. The proportional term acts on the immediate past value of the tracking error, integral terms provide memory effect by averaging the past errors, and, finally, derivative terms provides a prediction of the future errors. The effect of independently increasing the different tuning gains on the overall system performance is summarized in Table 2.1. It is to be noted here that depending on the systems being considered, various variants of PID controllers are used in practice. These variants are obtained by setting one or two terms in (2.1) to zero. Popular variants are PI and PD. Other variants are also possible. In the case of TST system, PI controller is more popular over PID. The derivative terms can amplify switching and measurement noises that are unavoidable for the power converters in TST. This motivated TST researchers to focus on PI over the PID controller.

2.1.1.1   Fractional order PID control

PID controller described by (2.1) uses integration and differentiation operation on the tracking error signal to generate the control signal. In the Laplace domain,

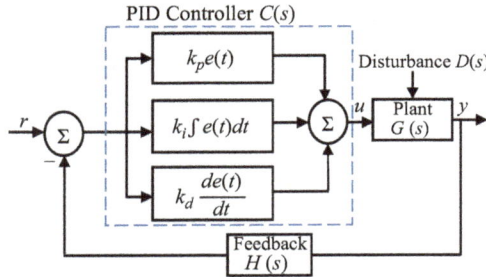

Figure 2.4 Overview of feedback control system with PID controller

Table 2.1 PID tuning gains effect on system characteristics

Tuning gain	Rise time	Overshoot	Settling time	Steady-state error	Stability
K_p	↓	↑	SC	↓	↓
K_i	↓	↑	↑	Eliminate	↓
K_d	SC	↓	↓	NE	Improve if small value

SC, small change; NE, no effect.

differentiation and integration are denoted by s and $1/s$. In integer order PID controller, the power of s is 1. In recent times, fractional order calculus became very popular and is used to develop fractional order PID (FOPID) controller, where the power of s is no longer an integer, rather it is a fractional number. This PID controller is also known as the $PI^\lambda D^\mu$ controller where the constant coefficients λ and μ are typically bounded as $0 < \lambda, \mu < 2$ [2,11]. It is clear to see that FOPID coincides with the conventional PID controller if the integration and differentiation order are selected as, $\lambda, \mu = 1$. Transfer function of the FOPID controller is given by:

$$C(s) = \frac{U(s)}{E(s)} = K_p + \frac{K_i}{s^\lambda} + K_d s^\mu \qquad (2.4)$$

A relationship between the FOPID and conventional PID controller is given in Figure 2.5.

Figure 2.5 shows that conventional PID (or variants) controller can have four distinct choices highlighted by the dot points. However, FOPID can have a large number of choices depending on the value of λ and μ. The additional choices give more design freedom, which leads to improvement in dynamic and steady-state responses compared to conventional PID controller. However, additional design freedom comes with additional tuning and implementation complexities. For a detailed description on the tuning of FOPID for TST, interested reader may consult [2] and the references therein.

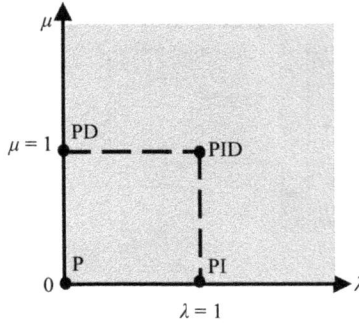

Figure 2.5   Relationship between integer and fractional order PID controllers

2.1.2   Sliding mode control

2.1.2.1   Basic concept of sliding mode control

The basic principle of SMC is to make any point in space reach the designed sliding mode surface (SMS) in limited time through switching control and ensure that the system reach the equilibrium point on the SMS. However, due to the inherent characteristics of discontinuous control and physical limitation of the actuator, it can cause small amplitude high-frequency up-down motion of the system states along the specified state trajectory. It is also known as "chattering". In order to understand the basic theory of SMC, sliding mode (SM) should be explained first. For this purpose, let us consider the following non-linear system:

$$\dot{x}(t) = f(x, u, t), x \in \mathbb{R}^n, u \in \mathbb{R}^m, t \in \mathbb{R} \tag{2.5}$$

where x, u, and t are the state variables of dimension n, control input of dimension m, and time, respectively. For this system, the following switching surface is considered:

$$s(x) = s(x_1, x_2, \ldots, x_n), s \in \mathbb{R}^n \tag{2.6}$$

In the state space, if there is $s(x) = 0$ in the hyperplane, this is the so-called SMS of the system. The main control objective is to make the system reach and maintain on this plane in a limited time. This surface is also known as the switching surface. Obviously, the state space will be divided into two parts $s(x) < 0$ and $s(x) > 0$ by SMS of the system $s(x) = 0$, as shown in Figure 2.6. On SMS of the state space, there are generally three cases of moving points:

- **Normal point:** Point A in Figure 2.6. When the moving point of the system state approaches near SMS of the system $s(x) = 0$, the moving point crosses SMS.
- **Starting point:** Point B in Figure 2.6. When the moving point of the system state approaches near SMS of the system $s(x) = 0$, it leaves from both sides of SMS.
- **Termination point:** Point C in Figure 2.6. The moving point of the system state approaches near SMS of the system $s(x) = 0$, the moving point crosses SMS from both sides.

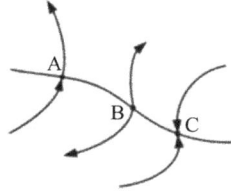

Figure 2.6 Motion state

Usually, the normal point and the starting point are meaningless, because their final motion directions are away from the SMS and they are difficult to be controlled on the SMS steadily. Consequently, SMC is mainly used to control the motion of the termination point. If all the moving points in a certain area on SMS are termination points like Point C, once the moving points approach this area, they will be attracted to move into this area. This area is also called the sliding mode band (SMB). All the motion in this band is SM motion. The existence of SMB is the sufficient conditions for SMC. In order to meet this requirement, when the motion points of the system move near SMS $s(x) = 0$, the following conditions must be satisfied [12]:

$$\lim_{s \to 0^+} \dot{s} < 0 \quad \text{and} \quad \lim_{s \to 0^-} \dot{s} > 0 \tag{2.7}$$

This means that in the neighborhood of SMS, the motion trajectory will reach SMS in a limited time. Therefore, it is also called local arrival condition. Furthermore, combining the above conditions, the inequality can be obtained as follows:

$$\lim_{s \to 0} s\dot{s} < 0 \tag{2.8}$$

According to the above inequality and Lyapunov stability theorem, the arrival condition of SM in the form of Lyapunov function can be expressed as [12]:

$$\dot{V}(x) < 0, V(x) = \frac{1}{2}s^2 \tag{2.9}$$

When the system in (2.5) satisfies the existence condition of SMB in (2.8), it is necessary to construct an appropriate control function to make the system slide on the SMS, which is the SMC law (SMCL). The control function $u(x)$ can be written as:

$$u(x) = \begin{cases} u^+(x), & s(x) > 0 \\ u^-(x), & s(x) < 0 \end{cases} \tag{2.10}$$

where $u^+(x)$ and $u^-(x)$ are smooth continuous function with $u^+(x) \neq u^-(x)$. Obviously, the form of control function on both sides of SMS $s(x) = 0$ is not the same. This is the special feature of SMC. When the system meets different conditions, it switches to different control structures. To achieve SMC, the following three conditions must to be satisfied [12]:

(i) Inequality (2.10) holds and the SM exists.
(ii) The arrival condition in (2.9) satisfies and the SMB exists.
(iii) SM motion of the system has good stability.

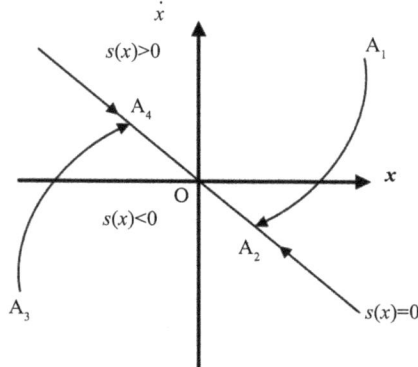

Figure 2.7   Sliding mode motion

In the system with SMC, the motion state is mainly divided into two stages as shown in Figure 2.7. When the switching function satisfies certain conditions, the trajectory of moving point A_1 (A_3) near the SMS is gradually transferred to point A_2 (A_4) in a limited time through the control function u(x). This trajectory is the approaching stage. After reaching SMS, the motion state of the system would be independent of the model, and the system would converge to the origin O, namely equilibrium point A_3 along SMS based on the sliding mode control law. This stage should be the sliding stage [12]. According to the SM motion state, the design of SMC is mainly divided into two steps: first, design a reasonable SMS for the system to make the system converge to the stable equilibrium point in a limited time after entering SMB and have good dynamic quality; second, design an appropriate SMCL in accordance with the arrival condition of SM to ensure the existence of SMB.

2.1.2.2   Types of sliding surfaces

The choice of sliding mode surface can be crucial in improving the performance of sliding mode controller. This has motivated researchers to propose different types of sliding mode surfaces. From the mathematical expression, it is mainly divided into three categories: namely linear, integral and terminal SMS.

(a)   **Linear sliding mode surface:** Since the beginning of SMC, linear sliding mode surface (LSMS) is the most widely used one due to its simplicity and easy implementation. A linear combination of system states and its derivative (or state tracking error) is often selected as the SMS, as seen in (2.11). The system has the characteristics of order reduction on LSMS:

$$s(x) = \sum_{i=1}^{n} C_i x_i \tag{2.11}$$

where C_i are positive real numbers and x_i are the state variables. The design of LSMS is simple, and it is suitable for both low-order and high-order systems.

By adjusting the coefficient/weight C_i, the desired dynamic response can be obtained. This makes the LSMS widely favored by the researchers. However, in any case, the system variables can only converge asymptotically and cannot reach the origin in a limited time. Moreover, most systems are not linear. With the increasing demand for fast and accurate industrial control system, the control of uncertain nonlinear systems have started to gain traction, where the usage of nonlinear SMS could be more useful compared to the linear counterpart.

(b) **Integral sliding mode surface:** In order to improve the performance of SMC, a variety of time-varying SMSs appear on the basis of LSMS. J-J Slotine and other researchers introduced an integral term into SMS, which is defined as the integral sliding mode surface (ISMS) [13,14]. Authors in [15] analyzed the robustness of ISMS and proposed an ISMS design method based on disturbance norm minimization. Based on ISMS in the form of what was proposed by Slotine *et al.*, a general form of ISMS was proposed in [16] to make the selection of ISMS parameters that have more degrees of freedom. SMC with this surface could reduce the chattering of the system by the integral action which would also decrease the steady-state error of the system. Thus, it has attracted the attention of many researchers and has been applied to numerous nonlinear systems including servo motor and manipulator. It has become one of widely used nonlinear SMSs. The common expression of this SMS is given by:

$$s(x) = k_p x + k_i \int_{-\infty}^{t} x(\tau)\, d\tau \tag{2.12}$$

where k_p and k_i are the positive gains. Initially Slotine [13] and later on Lee [17] proposed the whole process of ISMS which could make the system on SMS at the initial time to avoid the arrival process and have global robustness. Authors in [18] designed a nonlinear ISMS by a new saturation function with the function of "small error amplification and large error saturation." Authors in [19] proposed a SMS in the form of PID and applied it to the tracking control of manipulator. The results showed that the system with SMS in the form of PID had faster response speed than that in the form of PD.

(c) **Terminal sliding mode surface:** In 1988, Zak first proposed the concept of terminal attractor [20]. As the eigenvalue of the system is infinite at the origin, it can converge to the origin in a finite time. Venkataraman *et al.* applied it to the design of terminal sliding mode surface (TSMS) and proposed terminal sliding mode (TSM) control [21–23]. Although TSMS has fast convergence near the origin, its convergence speed is less than that of LSMS when the system states are far away from the origin. Therefore, Yu *et al.* proposed a fast terminal sliding mode surface (FTSMS), which combines the advantages of TSMS and LSMS [24]. Near the origin, TSMS plays a major role; when it is far from the origin, the system state has the dynamic performance of LSMS. The commonly used TSMS can be expressed as:

$$s(x) = \dot{x} + \beta x^{\frac{q}{p}} \tag{2.13}$$

where $\beta > 0$ and the positive odd numbers p and q satisfy the relationship $p > q$.

2.1.2.3 Basic concept of reaching law

The arrival condition of SM can only make the system reach SMS from any state point, and a reasonable SMS can only ensure the motion quality of the system in the SMB. However, these conditions cannot effectively control the motion trajectory of the system outside the SMB. Therefore, Gao *et al.* put forward the concept of reaching law for SMC [25]. There are four common sliding mode reaching laws and they are:

(i)   **Constant rate reaching law:**

$$\dot{s} = -\varepsilon \, \text{sign}(s), \varepsilon > 0 \tag{2.14}$$

where sign is the conventional signum function. This reaching law only guarantees that the approach speed is constant ε. The convergence time of the system moving to SMS can be accelerated or reduced by adjusting the gain term of the sign function ε. If ε is large, the reaching speed is large, which will inevitably increase the chattering and reduce the dynamic quality of the system; if ε is small, chattering can be reduced; however, it inevitably decreases the reaching speed to the SMS.

(ii)   **Constant plus proportional rate reaching law/exponential reaching law:**

$$\dot{s} = -\varepsilon \, \text{sign}(s) - \eta s, \qquad \varepsilon, \eta > 0 \tag{2.15}$$

The first-order linear function ηs is added to the exponential reaching law. This makes the system converge quickly when it is far from SMS, and approach at the constant speed ε close to SMS. In order to ensure the dynamic quality of the fast arrival process of SMS and reduce chattering, it is better to increase η and reduce ε at the same time.

(iii)   **Power rate reaching law:**

$$\dot{s} = -\varepsilon \, |s|^{\mu} \, \text{sign}(s), \qquad \varepsilon > 0, 0 < \mu < 1 \tag{2.16}$$

The reaching law is obtained by adding a power term on the basis of the constant rate reaching law, which can make the reaching rate of the moving point change adaptively with the distance between the system state variable and SMS. Through adjustment rate μ, when the system state is relatively distant from the SMS, it would approach SMB at a large speed; when the system state is near SMS, it approaches SMB at a small speed to reduce chattering

(iv)   **General reaching law:**

$$\dot{s} = -\varepsilon \, \text{sign}(s) - f(s), \qquad f(0) = 0, \text{ when } s \neq 0, \, sf(s) > 0 \tag{2.17}$$

In the general reaching law, a function $f(s)$ is added. The different suitable reaching law can be obtained by designing $f(s)$ according to the system requirement.

2.1.2.4 Chattering and solution

The sliding mode controller operates through switching. For an ideal SMC, the switching frequency of the system "structure" is infinite, the control force is infinite, and the system state measurements are accurate. In this way, SM is always a smooth motion with reduced dimension and asymptotically stable at the origin without chattering. However, for a real SMC, these assumptions cannot be fully established and the chattering is inevitable in practice. Moreover, if the chattering is eliminated, the ability of reject perturbation and disturbance is also disappeared. Therefore, the chattering can be only reduced to a certain extent and cannot be eliminated completely. The chattering becomes a prominent obstacle in the application of SMC in practical systems, and serious chattering will also lead to damaging consequences for the underlying system being controlled. Negative effects of chattering can be summarized as:

- It may affect the dynamic performance of the system and destroy the operating conditions of SM, resulting in excessive overshoot, longer transition process and even may make the system unstable in the long run.
- The chattering near the equilibrium point will decrease the static index of the system and will result in static error.
- It would cause mechanical wear and tear and increase control energy consumption.
- It may excite the strong oscillation of the unmodeled part of the system and become a vibration source, resulting in potentially damaging resonant motion.

Essentially, the chattering is caused by the nonlinearity of sliding mode control. The main causes of the chattering are summarized as follows:

(i) **Time delay of the switch**

   The ideal switch model sign(s) is always defined as follows:

$$\text{sign}(s) = \begin{cases} +1, & s > 0 \\ -1, & s < 0 \end{cases} \tag{2.18}$$

   Ideally, the system should be switched at $s = 0$. However, due to the time delay of the switch, the accurate control action to the state is delayed for a certain time τ. The typical time delay switch model is given in the following equation:

$$\text{Hys}^\tau(s) = \begin{cases} +1, & \text{when} \quad t < t_0 + \tau, \dot{s} < 0 \quad \text{or} \quad t > t_0 + \tau, \dot{s} > 0 \\ -1 & \text{when} \quad t < t_0 + \tau, \dot{s} > 0 \quad \text{or} \quad t > t_0 + \tau, \dot{s} < 0 \end{cases} \tag{2.19}$$

   As the amplitude of the control quantity decreases gradually with the amplitude of the state quantity, it is represented by superimposing an attenuated triangular wave on the smooth SMS.

(ii) **Spatial lag of the switch**

   The spatial lag effect of the switch is equivalent to the existence of a "dead zone" in the state space. The relative formula can be described by (2.20), where an equal amplitude waveform is superimposed on the smooth SMS:

$$\text{Hys}^\tau(s) = \begin{cases} +1, & \text{when} \quad s < -\Delta \quad \text{or} \quad |s| < \Delta, \dot{s} > 0 \\ -1 & \text{when} \quad s > +\Delta \quad \text{or} \quad |s| < \Delta, \dot{s} < 0 \end{cases} \tag{2.20}$$

(iii) **Inertia of the system**

   As the control force of the system cannot be infinite, it will undoubtedly limit the acceleration of the system. The inertia of the system always exists which make the switching of control lag. The chattering caused by this reason is similar to that caused by time delay of the switch.

(iv) **Error of state measurement**

The state measurement error mainly perturbs the SMS and is often accompanied by randomness coming from the various noise sources. Therefore, the chattering presents irregular attenuated triangular wave. The bigger the measurement error, the greater the chattering amplitude.

(v)   **Discrete system**

The SM of discrete system is "quasi-SM". Its switching action does not happen on the SMS, but on the surface of a cone with the origin as the vertex. Therefore, there is attenuated chattering. The larger the cone, the greater the chattering amplitude. The size of the cone is related to the sampling period.

In summary, when the trajectory of the system reaches SMS, the system inertia makes the moving point pass through SMS, resulting in chattering on the ideal SM. The strength of this chattering is related to all the above factors. In high-spec industrial control computer, the high-speed logic conversion and high-precision numerical operation make the influence of the time delay and spatial lag of the switch almost nonexistent. Therefore, the control discontinuity caused by the switching action is the essential cause of the chattering in this case.

2.1.2.5   Research on chattering reduction

The chattering suppression is not only a theoretical problem but also an engineering problem. Many researchers focus on the anti-chattering of SMC due to the possible serious consequences of the chattering. At present, these methods mainly include:

(a)   **Quasi-SMC**

In the 1980s, J-J Slotine and colleagues proposed the concepts of "Quasi-SM" and "boundary layer" into SMC [26]. The so-called Quasi-SM refers to the mode in which the motion trajectory of the system is limited to a neighborhood δ of the ideal SM. This neighborhood is usually called the "boundary layer" of SMS. The smaller the thickness of the boundary layer, the better the control performance. However, it will increase the control gain and enhance the chattering. On the contrary, the greater the thickness of the boundary layer, the smaller the chattering. Nevertheless, it will deteriorate the control performance. For the Quasi-SMC, normal SMC was used outside the boundary layer, and the continuous feedback control was selected in the boundary layer. It will effectively avoid or weaken the chattering. In the boundary layer, several switching functions are proposed to replace the signum function sign(s) for chattering suppression. Commonly used functions are as follows:

• Saturation function

$$\text{sat}(s) = \begin{cases} +1, & s > \delta \\ \dfrac{s}{\delta}, & |s| \leq \delta \\ -1, & s < -\delta \end{cases} \tag{2.21}$$

- Continuous function

$$\Theta(s) = \frac{s}{|s| + \delta} \tag{2.22}$$

- Hyperbolic tangent function

$$\tanh\left(\frac{s}{\delta}\right) = \frac{s^{\frac{s}{\delta}} - s^{-\frac{s}{\delta}}}{s^{\frac{s}{\delta}} + s^{-\frac{s}{\delta}}} \tag{2.23}$$

(b) **Reaching law**

The reason for the chattering is that the moving point of the system state has a large speed when it rushes to SMS with its inherent inertia. Therefore, the above reaching laws can be selected and designed for the moving point. By designing a reasonable arrival speed, the chattering of the system can be reduced.

(c) **Observer method**

In the conventional SMC, a large switching gain is often required to eliminate the external disturbance and uncertainty, which are the main sources of chattering. According to the measurement results of the observer for the external disturbance and uncertainty, the switching gain can be designed to reduce the effect of chattering.

(d) **Filter method**

Filter is an effective method to eliminate high-frequency signal. Therefore, low-pass filter can be used to smooth the output signal of SMC, which is also an effective method to eliminate high-frequency chattering.

(e) **Intelligent control**

In the literature, intelligent control methods e.g. fuzzy algorithm, neural network, genetic algorithm, etc. are combined with the SMC to reduce the chattering. Thanks to the self-learning, self-adaptation, self-organization, and evolution aspect of the intelligent control, the gain of switching item in SMC can be automatically adjusted online according to the current system state, so as to minimize the influence of switching action and realize the online optimization of the controller output. This is also a useful method to eliminate chattering in SMC.

(f) **Other methods**

In addition to the above methods, some other methods e.g. reducing switching gain, sliding sector method, dynamic SMC, etc. are also available in the literature.

The above methods can suppress the chattering of SMC to a certain extent. However, they lose the remarkable advantage of the insensitivity to the internal parameter perturbation and external disturbance more or less. The high order sliding mode control (HOSMC) could retain this insensitivity and reduce the chattering simultaneously. However, this comes at the cost of additional computational complexity resulting from the system relative order increase.

2.1.2.6 Brief overview of HOSMC

In order to explain HOSMC, the sliding order r should be introduced first, which is a very important index to distinguish different SMC algorithms. The sliding order

r refers to the number of sliding mode variables s and all its continuous derivatives (including zero order) equal to 0 on SMS $s = 0$. The r order sliding set of SMS $s = 0$ is described by (2.24), which constitutes the r-dimensional constraints of the dynamic system state:

$$s = \dot{s} = \ddot{s} = \ldots = s^{(r-1)} = 0 \tag{2.24}$$

If the r order sliding set in (2.24) is non-empty and it is assumed to be a local integral set in the sense of Filippov (i.e., it consists of Filippov trajectories of discontinuous dynamic systems), the relevant motion satisfying equation (2.24) is called "r order sliding mode" with respect to SMS $s = 0$ [27]. This parameter describes the dynamic smoothness of the system constrained on SMS. According to the above definition, the sliding order r of the conventional SM is 1 as $\dot{s}$ is discontinuous on SMS $s = 0$. The conventional SMC is also called the first-order SMC (FOSMC).

The restriction of the relative order of SM and the chattering of FOSMC severely limit its application. To overcome this problem, Levant systematically proposed the concept of HOSMC for the first time in his doctoral thesis in 1987 [28]. He pointed out that HOSMC could essentially avoid the relative order restriction in FOSMC, weaken the chattering, increase the control accuracy, and retain the advantages of FOSMC. Since then, Emelyanov, Levant, and many other researchers proposed different HOSMC algorithms at the early stage, including twisting algorithm and super-twisting algorithm, and gave detailed algorithmic analysis [29,30].

In the early 1990s, although the theoretical foundation of twisting and super-twisting-based HOSMC have been formed, the theory has not attracted much attention due to some theoretical defects. In 1994, Fridman and Levant introduced the theory of HOSMC [27]. Then, the theory has finally received the much deserved attention, and many researchers began to focus on the theory and application of HOSMC. The discontinuous term of the control output of HOSMC exists in the integral term. Therefore, the continuous smooth signal in time is obtained. It could eliminate chattering theoretically and still retain the strong robustness of FOSMC. The relative research shows that HOSMC has better comprehensive performance than FOSMC. The control accuracy of HOSMC is higher in the steady state, especially in the discrete-time control system.

2.1.2.7   Basic concept of second-order SMC (SOSMC)

HOSMC mainly includes second-order SMC (SOSMC) and arbitrary order SMC higher than second order. As arbitrary order SMCs require more information and the relative structure is more complicated, SOSMC is the most widely used HOSMC at present. According to the definition, in the system state space, if and only if the system trajectory is located at the junction of plane $s = 0$ and $\dot{s} = 0$, the system has second-order sliding mode (SOSM) ($r = 2$), as shown in Figure 2.8.

Consider the single input dynamical system as follows:

$$\dot{x} = f(t,x) + g(t,x)u, \quad s = s(t,x) \tag{2.25}$$

where $x \in \mathbb{R}^n$ are the state variables, $u \in \mathbb{R}$ is the control input, $f(t,x)$, $g(t,x)$ are unknown smooth functions, and $s(t,x) = 0$ is the SMS of the system. Obviously,

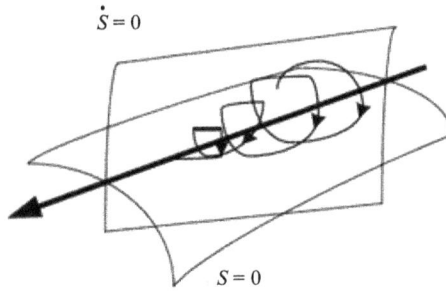

Figure 2.8 The trajectory of SOSM

the control objective of SOSMC is to make the system (2.25) converge to SMS in a finite time and have SOSM. It means that the system state reaches the following SM manifold in a finite time:

$$s(t,x) = \dot{s}(t,x) = 0 \tag{2.26}$$

Set the dummy variable $x_{n+1} = t$ and extend the system (2.25) by considering $x_e = \left(x^T, x_{n+1}\right)^T$, $f_e = \left(f^T, 1\right)^T$, $g_e = \left(g^T, 0\right)^T$. Then, the extended system can be written as:

$$\dot{x}_e = f_e\left(x_e\right) + g\left(x_e\right)u, \quad s = s\left(x_e\right) \tag{2.27}$$

According to the definition of r sliding order, the following two different cases are considered for the sliding mode variable s in SOSMC:

(i) Relative order $r = 1$, $\dfrac{\partial \dot{s}}{\partial u} \neq 0$

(ii) Relative order $r = 1$, $\dfrac{\partial \dot{s}}{\partial u} = 0$, $\dfrac{\partial \ddot{s}}{\partial u} \neq 0$

For the first case, FOSMC can be used to solve the control problem, but with some chattering. If SOSMC is used, the chattering can be effectively reduced. At this time, the derivative of the control input u is regarded as the new control variable. Then, the discontinuous control variable designed to make the sliding mode variable s converge to zero and maintain SOSM, that is, $s = \dot{s} = 0$. The control input u is obtained by integrating $\dot{u}$. Therefore, this control is continuous and the chattering can be suppressed under the action of integration theoretically. The first derivative $\dot{s}$ of sliding mode variable s is obtained as follows:

$$\dot{s} = \frac{\partial s}{\partial x}\left[f_e\left(x_e\right) + g_e\left(x_e\right)u\right]L_{f_e}s + L_{g_e}su \tag{2.28}$$

where $L_{f_e}s = \dfrac{\partial s}{\partial x_e}f_e(x_e)$ and $L_{g_e}s = \dfrac{\partial s}{\partial x_e}g_e(x_e)$ are Lie derivatives of s with respect to or along f_e and g_e, respectively [31]. The second derivative $\ddot{s}$ of sliding mode variable s can be obtained as follows:

$$\begin{aligned}
\ddot{s} &= \frac{\partial \left(L_{f_e}s + L_{g_e}su\right)}{\partial x_e}\left[f_e(x_e) + g_e(x_e)u\right] + \frac{\partial \dot{s}}{\partial u}\dot{u} \\
&= L_{f_e}^2 s + L_{f_e}L_{g_e}su + L_{g_e}L_{f_e}su + L_{g_e}^2 su^2 + L_{g_e}s\dot{u} \\
&= \varphi_A(t,x,u) + \gamma_A(t,x)\dot{u}(t)
\end{aligned} \qquad (2.29)$$

where $\varphi_A(t,x,u) = L_{f_e}^2 s + L_{f_e}L_{g_e}su + L_{g_e}L_{f_e}su + L_{g_e}^2 su^2 = \ddot{s}\,|_{\dot{u}=0}$ and $\gamma_A(t,x) = L_{g_e}s = \dfrac{\partial \ddot{s}}{\partial \dot{u}} \neq 0$. From (2.29), the control input u can be regarded as the uncertain interference term of φ_A and the derivative $\dot{u}$ of the control input u is the new control variable to be designed.

For the second case, the control input u does not directly affect the dynamic characteristics of $\dot{s}$, but directly affects the dynamic characteristics of $\ddot{s}$. Equation (2.29) can be rewritten as follows:

$$\begin{aligned}
\ddot{s} &= \frac{\partial \left(L_{f_e}s + L_{g_e}su\right)}{\partial x_e}\left[f_e(x_e) + g_e(x_e)u\right] + \frac{\partial \dot{s}}{\partial u}\dot{u} \\
&= L_{f_e}^2 s + L_{g_e}L_{f_e}su \\
&= \varphi_B(t,x,u) + \gamma_B(t,x)u(t)
\end{aligned} \qquad (2.30)$$

where $\varphi_B(t,x,u) = \ddot{s}\,|_{u=0} = L_{f_e}^2 s$ and $\gamma_B(t,x) = \dfrac{\partial \ddot{s}}{\partial u} = L_{g_e}L_{f_e}s \neq 0$. According to (2.30), the relative order of the sliding mode variable s with respect to the control input u is 2. In this case, the control output u is discontinuous, which is unlike the previous one.

These two cases can be considered for a class of second-order uncertain affine nonlinear system. In the first case, the relevant control signal is the derivative $\dot{u}$ of the actual control input u; in the second case, the control signal is the actual control input u. Therefore, the above two cases of SOSMC problem can be all transformed into the finite time stabilization problem of the following nonlinear systems:

$$\begin{aligned}
\dot{y}_1(t) &= y_2(t) \\
\dot{y}_2(t) &= \varphi(t,x) + \gamma(t,x)\xi(t) \\
s(t,x) &= y_1(t)
\end{aligned} \qquad (2.31)$$

For (2.31), $\varphi(t,x)$, $\gamma(t,x)$, and $\xi(t)$ would have different meanings and structures in the two different cases. In order to meet the robustness of SOSMC to uncertainty, it is necessary to assume the global boundedness of uncertain disturbances [32]:

$$\begin{cases} |\varphi| \leq C \\ 0 < K_m \leq \gamma \leq K_M \end{cases} \qquad (2.32)$$

where C, K_m, and K_M are positive real numbers.

2.1.2.8 Basic algorithms of SOSMC

Presently, the commonly used SOSMC algorithms include twisting algorithm, sub-optimal algorithm, prescribed convergence law algorithm and super-twisting algorithm. A brief overview of these algorithms is presented below:

- **Twisting algorithm**

  Twisting algorithm is the first proposed SOSMC algorithm [29,30]. The mathematical expression is as follows:

$$u = -r_1 \mathrm{sign}(s) - r_2 \mathrm{sign}(\dot{s}), \quad r_1, r_2 > 0 \tag{2.33}$$

  The sufficient conditions for its finite time stability are:

$$(r_1 + r_2) K_m - C > (r_1 - r_2) K_M + C, \quad (r_1 - r_2) K_m > C \tag{2.34}$$

  According to the above formula, the design of this algorithm needs the time derivative of sliding mode variable $\dot{s}$. Under this control algorithm, the phase trajectory of the system rotates around the origin, as shown in Figure 2.9. Precisely speaking, the phase trajectory can converge to the origin after infinite spiral motions in a finite time. In accordance with the relevant literature, the absolute value of the intersection between the phase trajectory and the coordinate axis decreases in the form of proportional series with the number of rotations.

- **Sub-optimal algorithm**

  This algorithm was proposed by Bartolini in [33,34]. The mathematical expression can be summarized as follows:

$$u = -r_1 \mathrm{sign}\left(s - \frac{s^\star}{2}\right) + r_2 \mathrm{sign}(s^\star), \quad r_1, r_2 > 0 \tag{2.35}$$

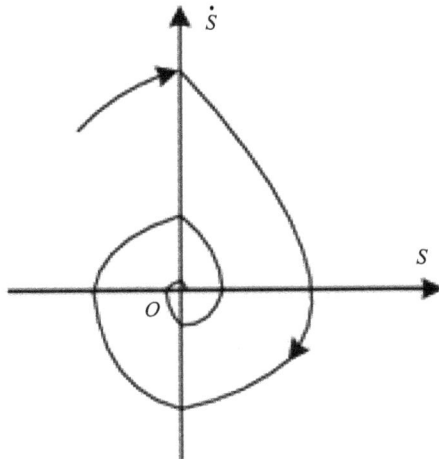

Figure 2.9   The phase trajectory under twisting algorithm on s–$\dot{s}$ plane

where s^* represents the value of s at the last moment when $\dot{s} = 0$. The initial value of s^* is 0. According to [33,34], a concise sufficient condition for the finite time convergence of sub-optimal algorithm is as follows:

$$\begin{cases} 2\left[(r_1 + r_2)\,K_m - C\right] > C + (r_1 - r_2)\,K_M \\ (r_1 - r_2)\,K_m > C \end{cases} \tag{2.36}$$

In fact, the sub-optimal algorithm is evolved from the time optimal controller of the double integrator system and the relative convergence region can be set in advance. The phase trajectory on s–$\dot{s}$ plane is limited to a finite parabola containing the origin, as shown in Figure 2.10. The precondition of this algorithm is that s and its derivative $\dot{s}$ are known.

- **Prescribed convergence law algorithm** The common expression of prescribed convergence law algorithm is as follows [30,32]:

$$u = -\alpha \operatorname{sign}\left[\dot{s} + \beta\,|s|^{\frac{1}{2}}\operatorname{sign}(s)\right], \alpha, \beta > 0 \tag{2.37}$$

The sufficient condition to ensure the convergence of the system in finite time is defined in the following equation:

$$\alpha K_m - C > \frac{\beta^2}{2} \tag{2.38}$$

Prescribed convergence law algorithm essentially draws lessons from the advantages of the conventional SMC and constructs the nonlinear SMS $s = \dot{s} + \beta\,|s|^{\frac{1}{2}}\operatorname{sign}(s) = 0$. On s–$\dot{s}$ plane, the system phase trajectory will reach the predetermined SMS and gradually converge to the origin along SMS in finite time. However, it is difficult to estimate the convergence time. The phase trajectory is shown in Figure 2.11.

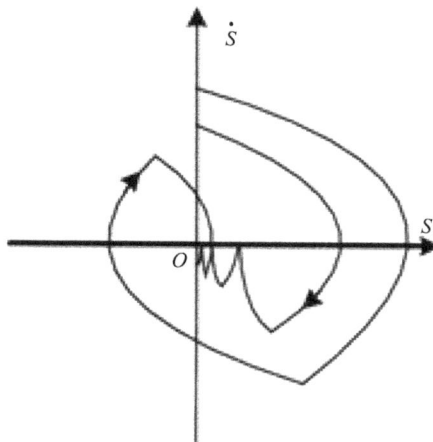

Figure 2.10   The phase trajectory under sub-optimal algorithm on s–ṡ plane

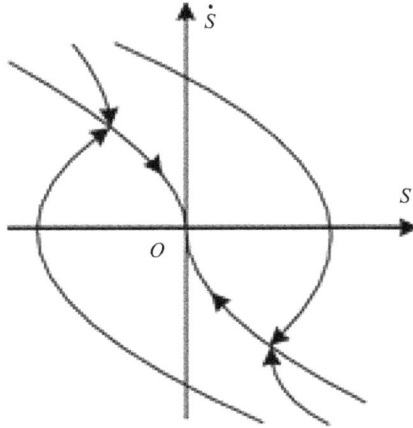

Figure 2.11   The phase trajectory under prescribed convergence law algorithm on s–ṡ plane

- **Super-twisting algorithm**
  Compared to the previous three algorithms, super-twisting algorithm has two important characteristics:
  (i) Unlike the conventional SOSMC, it only needs the information of the sliding mode variable s and does not need to know the information of $\dot{s}$;
  (ii) It is a SOSMC that could be directly applied to the system with the relative order of 1 and does not need to introduce a new control variable.
  The algorithm form of super-twisting algorithm, which is proposed by Levant, is as follows [30,32]:

$$u = -\beta \, |s|^{\frac{1}{2}} \, \text{sign}(s) + u_1$$
$$\dot{u}_1 = -\alpha \, \text{sign}(s)$$

(2.39)

If the system (2.27) could satisfy the following condition:

$$\left| L_{fe}^2 s + L_{fe} L_{ge} su + L_{ge} L_{fe} su + L_{ge} su^2 \right| \le C, \ 0 < K_m \le L_{ge} s \le K_M$$

(2.40)

The sufficient conditions for its finite time stability are given by [32]:

$$\begin{cases} \alpha > \dfrac{C}{K_m} \\ \beta^2 > 2\dfrac{\alpha K_M + C}{K_M} \end{cases}$$

(2.41)

From these two inequalities, it can be seen that the controller parameters should also be selected according to the uncertainty bound for finite-time stability. The phase trajectory of super-twisting algorithm is shown in Figure 2.12.

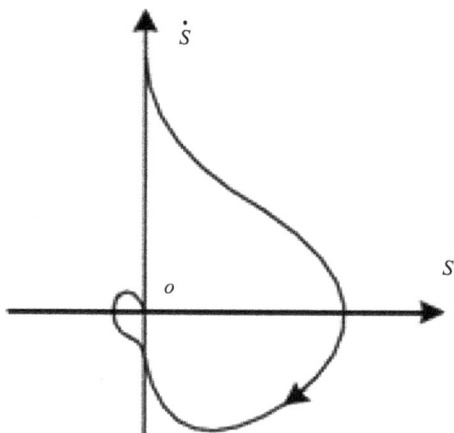

Figure 2.12 The phase trajectory under super-twisting algorithm on s–ṡ plane

2.2 Mathematical modelling

In this section, mathematical models of various components of the TST system are presented. These models are essential for simulation and control design purpose.

2.2.1 Tidal resource model

Tidal stream current is periodically in motion. In fact, it is not easy to get the exact movement. In any hydrodynamic model for the current flow in a channel, there is a requirement for accurate water height data. For any subsequent resource evaluation and site capacity estimation, there must be a large amount of data available (usually at least 1 year). There are several methods to model the marine current such as first order model – SHOM (service hydrographique et océanographique de la marine), harmonic analysis method (HAM), double cosine method, tidesim, tide 2D, etc. Here the first-order model is presented. More information on the other methods can be found in [6,35–37]. The SHOM, which is a first-order model, needs the current velocities for spring and neap tides [6]. These values should be given at hourly intervals starting at 6 h before high waters and ending 6 h after. Therefore, knowing tide coefficients, it is easy to derive a simple and practical first-order model for tidal current velocity V_{tide} and is given by:

$$V_{tide} = V_{nt} + \frac{(C - 45)(V_{st} - V_{nt})}{95 - 45} \tag{2.42}$$

where V_{nt} is the hourly intervals neap tide current velocities, V_{st} is the hourly intervals spring tide current velocities, and C is the tide coefficient which is defined by astronomic calculation of earth and moon position. In (2.42), 95 and 45 are, respectively, mean spring and neap tide medium coefficients.

2.2.2 *Tidal turbine model*

Like wind energy, regardless of turbine design, only a small portion of the hydrody-namic energy can be absorbed from the water. The amount of extractable mechanical power P_m is expressed by the following formula [8,38,39]:

$$P_m = \frac{1}{2} C_p \rho A V_{tide}^3 \tag{2.43}$$

where ρ is the density of the ocean water, A is the cross-section of TST turbine with $A = \pi R^2$, R is the radius of the turbine, V_{tide} is the tidal velocity and C_p is the power coefficient, which is also called Betz's coefficient. The Betz's coefficient C_p highly depends on the tip speed ratio (TSR) denoted by λ, the pitch angle β and the number and the geometry of the blades. More precisely, it is determined by the drag ("D") and lift ("L") forces on the blade. These two forces depend on many factors, such as the seawater density ρ, the resultant velocity "W" (m/s), the attack angle "α" (rad) of the water flow, the blade chord ("r") (m), the drag and lift coefficients "$C_D(\alpha)$," and "$C_L(\alpha)$." The local drag and lift gradients "dD" & "dL" on the blade section between radius r and elementary radius $(r + dr)$ are defined as:

$$\frac{dD}{dr} = \frac{1}{2}\rho C_D(\alpha) W^2 \text{chord}(r)$$

$$\frac{dL}{dr} = \frac{1}{2}\rho C_L(\alpha) W^2 \text{chord}(r) \tag{2.44}$$

A simple overview of the blade section dynamics is shown in Figure 2.13. According to Figure 2.13, the resultant velocity "W" is defined as [40]:

$$W^2 = \left\{ V_{tide}(1-a)^2 \right\} + \left\{ \omega r(1+b) \right\}^2 \tag{2.45}$$

According to the blade element momentum theory, the relation between the C_p gradients and λ can be easily achieved as shown below [40]:

$$\frac{dC_p}{dr} = \frac{\lambda(1-a)^2 \text{chord}(r) N_b r C_{F_T}}{R^3 \pi \sin^2(\phi)} \tag{2.46}$$

$$C_{F_T} = C_L(\alpha)\sin(\phi) - C_D(\alpha)\cos(\phi)$$

$$C_{F_N} = C_L(\alpha)\cos(\phi) + C_D(\alpha)\sin(\phi) \tag{2.47}$$

The theoretical maximum C_p is 16/27 (0.59259). For TST, the maximum C_p is around in the range 0.35–0.5 [1]. The TSR (λ) is defined as:

$$\lambda = \frac{R\omega}{V_{tide}} \tag{2.48}$$

where ω is the speed of the turbine. According to this theory, the turbine rotor model was validated through comparing the simulation model with experimental data from the available literature [41–43] (cf. Figure 2.14). The obtained power coefficient C_p and the extractable power curves are shown in Figure 2.15. The speed references based on maximum power point tracking (MPPT) strategy are shown in Figure 2.16 [6].

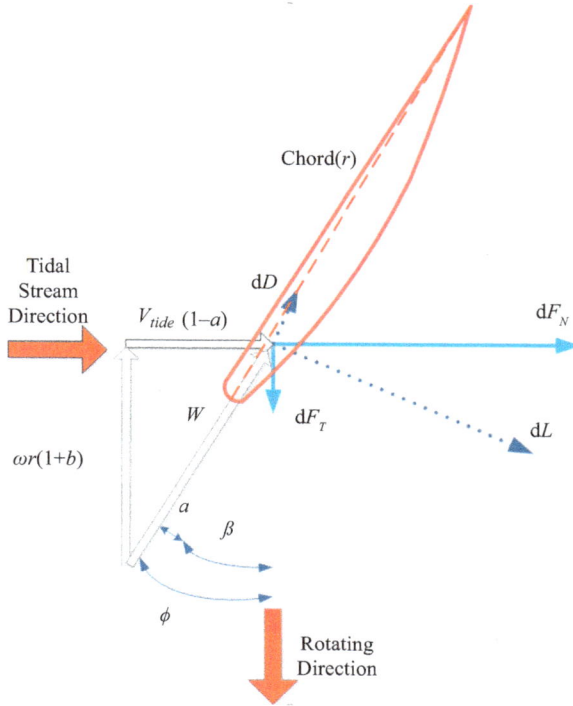

Figure 2.13   Blade section dynamics

Figure 2.14   The tested turbine [41,42]

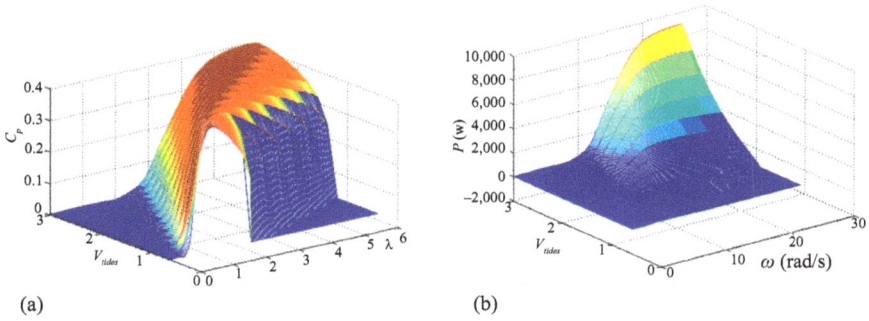

(a)   (b)

Figure 2.15   Turbine performance [6]. (a) C_p (λ, V_{tide}) curves and (b) the extractable power P_m (ω, V_{tide}).

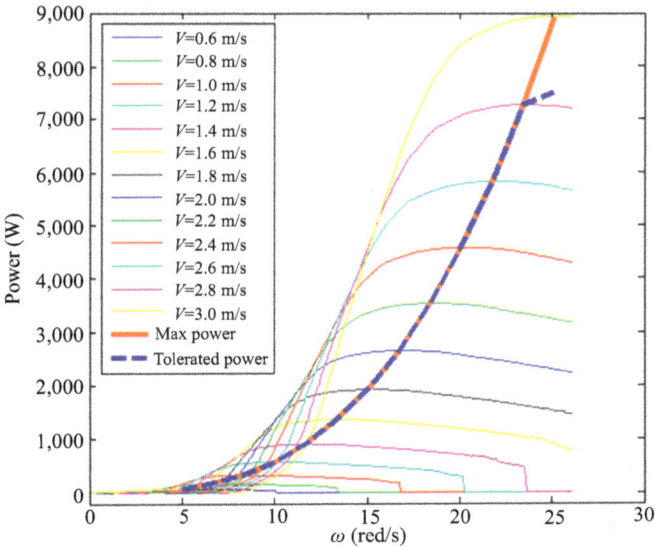

Figure 2.16   Power curves for different tidal current speed [6]

2.2.3   Modeling of the generator

Right now, much of the technology which has been suggested for the extraction of tidal current energy is the same as that used for wind energy applications. It is obvious that some wind electrical generator topologies could be used for TST [44]. Normally, the major generator topologies are: induction generator (IG), synchronous generator (SG), doubly-fed induction generator (DFIG), and permanent magnet synchronous generator (PMSG). Although PMSG and IG seem to be the most interesting choices in TST as many projects adopt these two topologies, according to the successful

Table 2.2   Generator topologies comparisons [1]

	Pros	Cons
IG	✓ Full speed range ✓ No brushes on the generator ✓ Complete control of reactive and active power ✓ Proven technology	✗ Full scale power converter ✗ Need for gear
SG	✓ Full speed range ✓ Possible to avoid gear ✓ Complete control of reactive and active power	✗ Small converter for field ✗ Full scale power converter ✗ Multipole generator (big and heavy) in case of direct driven topology
PMSG	✓ Full speed range ✓ Possible to avoid gear ✓ Complete control of reactive and active power ✓ Brushless (low maintenance) ✓ No power converter for field	✗ Full scale power ✗ converter ✗ Multipole generator (big and heavy) ✗ Permanent magnets needed
DFIG	✓ Limited speed range −30–30% around synchronous speed ✓ Inexpensive small capacity PWM inverter ✓ Complete control of reactive and active power	✗ Need slip rings ✗ Need for gear

cases of wind turbine, PMSG and DFIG could be the most suitable topologies for large capacity TSTs [45]. The pros and cons of these four generator topologies are summarized in Table 2.2.

Beside these conventional generator topologies, some other researchers also proposed some non-conventional generator topologies for TST, such as: Doubly Salient Permanent Magnet Generator (DSPMG), Double Stator Cup Rotor Permanent Magnet Generator (DSCRPMG), Double-Stator Axial Flux Permanent Magnet Generator (DSAFPMG), etc. [45–47]. Consequently, in the following part, the modeling of PMSG, DFIG and DSPMG is presented first; then, SMC is applied in these three generators-based TST.

2.2.4   Modeling of DFIG

Although DFIG has been already widely used in wind turbine applications, there is no relative large-scale project claiming to use DFIG in practice. As the maximum capacity of DFIG has reached MW-scale, DFIG could be considered as an interesting candidate for even moderately large TST applications. A schematic diagram of

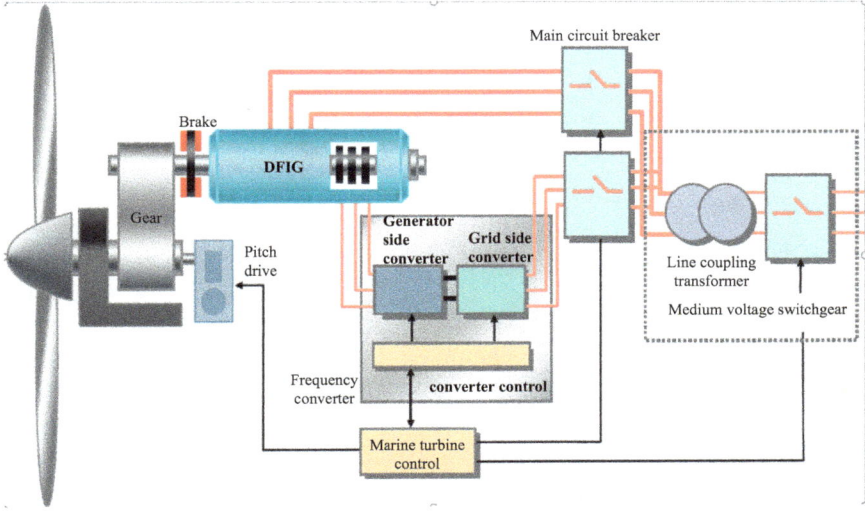

Figure 2.17   Schematic diagram of DFIG-based TST [6]

DFIG-based TST is shown in Figure 2.17 [6]. The stator and rotor voltage equations in the stationary reference frame are given by:

$$[v_s]_3 = R_s [i_s]_3 + \frac{d}{dt} [\phi_s]_3$$

$$[v_r]_3 = R_r [i_r]_3 + \frac{d}{dt} [\phi_r]_3$$

(2.49)

where the subscript s and r indicate stator and rotor, $[v_s]_3$ and $[v_r]_3$ are the stator voltage and rotor voltage vectors with $[v_s]_3 = [v_{sa}\ v_{sb}\ v_{sc}]^T$, $[v_r]_3 = [v_{ra}\ v_{rb}\ v_{rc}]^T$; R_s and R_r are the stator resistance and rotor resistance; $[i_s]_3$, $[i_r]_3$ are the stator current and rotor current vectors with $[i_s]_3 = [i_{sa}\ i_{sb}\ i_{sc}]^T$, $[i_r]_3 = [i_{ra}\ i_{rb}\ i_{rc}]^T$; $[\phi_s]_3$, $[\phi_r]_3$ are the stator flux and rotor flux vectors with $[\phi_s]_3 = [\phi_{sa}\ \phi_{sb}\ \phi_{sc}]^T$, $[\phi_r]_3 = [\phi_{ra}\ \phi_{rb}\ \phi_{rc}]^T$. The stator and rotor flux are defined as:

$$[\phi_s]_3 = [L_{ss}] [i_s]_3 + [M_{sr}] [i_r]_3$$

$$[\phi_r]_3 = [L_{rr}] [i_r]_3 + [M_{rs}] [i_s]_3$$

(2.50)

with

$$[L_{ss}] = \begin{bmatrix} L_{ms} + L_{ls} & -\frac{1}{2}L_{ms} & -\frac{1}{2}L_{ms} \\ -\frac{1}{2}L_{ms} & L_{ms} + L_{ls} & -\frac{1}{2}L_{ms} \\ -\frac{1}{2}L_{ms} & -\frac{1}{2}L_{ms} & L_{ms} + L_{ls} \end{bmatrix}$$

$$
[L_{rr}] = \begin{bmatrix} L_{mr} + L_{lr} & -\dfrac{1}{2}L_{mr} & -\dfrac{1}{2}L_{mr} \\[2mm] -\dfrac{1}{2}L_{mr} & L_{mr} + L_{lr} & -\dfrac{1}{2}L_{mr} \\[2mm] -\dfrac{1}{2}L_{mr} & -\dfrac{1}{2}L_{mr} & L_{mr} + L_{lr} \end{bmatrix}
$$

$$
[M_{sr}] = [M_{rs}] = L_{ms} \begin{bmatrix} \cos(\theta_r) & \cos\left(\theta_r - \dfrac{2\pi}{3}\right) & \cos\left(\theta_r + \dfrac{2\pi}{3}\right) \\[2mm] \cos\left(\theta_r + \dfrac{2\pi}{3}\right) & \cos(\theta_r) & \cos\left(\theta_r - \dfrac{2\pi}{3}\right) \\[2mm] \cos\left(\theta_r - \dfrac{2\pi}{3}\right) & \cos\left(\theta_r + \dfrac{2\pi}{3}\right) & \cos(\theta_r) \end{bmatrix}
$$

where $[L_{ss}]$ and $[L_{rr}]$ are the stator and rotor inductance matrices, $[M_{sr}]$ and $[M_{rs}]$ are the mutual inductance matrices between the stator and rotor, L_{ms} and L_{mr} are the stator/rotor mutual inductances corresponding to the maximum flux of one stator/rotor phase winding, typically $L_{ms} = L_{mr}$, L_{ls} and L_{lr} are stator/rotor leakage inductances, and θ_r is the rotor electrical angle. To control the active and reactive power without coupling, DFIG is generally defined in the synchronous reference frame, i.e., $d - q$. The Park transformation for the stator and rotor is given by:

$$
C_{(s,r)abc \rightarrow dq} = \frac{2}{3} \begin{bmatrix} \cos\left(-\theta_{(s,sl)}\right) & \cos\left(-\theta_{(s,sl)} + \dfrac{2\pi}{3}\right) \\[2mm] \sin\left(-\theta_{(s,sl)}\right) & \sin\left(-\theta_{(s,sl)} + \dfrac{2\pi}{3}\right) \end{bmatrix}
$$

$$
\left. \begin{matrix} \cos\left(-\theta_{(s,sl)} - \dfrac{2\pi}{3}\right) \\[2mm] \sin\left(-\theta_{(s,sl)} - \dfrac{2\pi}{3}\right) \end{matrix} \right]
$$

$$(2.51)$$

where $\theta_{(s,sl)}$ are the stator synchronous rotating electrical angular and the slip electrical angular relative to the stator synchronous rotating electrical angular θ_s separately. The model in the $d - q$ coordinate axis of the stator is described by:

$$
v_{sd} = R_s i_{sd} + \frac{d}{dt}\phi_{sd} - \omega_s \phi_{sq}
$$

$$
v_{sq} = R_s i_{sq} + \frac{d}{dt}\phi_{sq} + \omega_s \phi_{sd}
$$

$$(2.52)$$

$$
v_{rd} = R_r i_{rd} + \frac{d}{dt}\phi_{rd} - (\omega_s - \omega_r)\phi_{rq}
$$

$$
v_{rq} = R_r i_{rq} + \frac{d}{dt}\phi_{rq} + (\omega_s - \omega_r)\phi_{rd}
$$

$$\phi_{sd} = L_s i_{sd} + L_m i_{rd}$$

$$\phi_{sq} = L_s i_{sq} + L_m i_{rq}$$

$$\phi_{rd} = L_r i_{rd} + L_m i_{sd} \tag{2.53}$$

$$\phi_{rq} = L_r i_{rq} + L_m i_{sq}$$

where v_{sd}, v_{sq}, v_{rd}, v_{rq}, i_{sd}, i_{sq}, i_{rd}, i_{rd}, and ϕ_{sd}, ϕ_{sq}, ϕ_{rd}, ϕ_{rq} are the voltage/current/flux component of the stator/rotor in the synchronously $d - q$ frame, respectively; ω_s and ω_r are stator synchronous electrical speed and rotor rotating electrical speed separately; L_m is the mutual inductance between the coaxial equivalent windings in $d - q$ coordinate system of the stator and rotor, $L_m = 1.5 L_{ms}$; L_s is the self-inductance of the equivalent two-phase winding of the stator in the $d - q$ coordinate system, $L_s = L_{lm} + L_{ls}$; L_r is the self-inductance of the equivalent two-phase winding of the rotor in the $d - q$ coordinate system and $L_r = L_m + L_{lr}$.

The electromagnetic torque (T_{em}) of DFIG is given by:

$$T_{em} = \frac{3}{2} n_p L_m \left(i_{sq} i_{rd} - i_{rq} i_{sd} \right) \tag{2.54}$$

where n_p is the pole pair. The mechanical equation can be written as:

$$J \frac{d\omega_m}{dt} = T_m - T_{em} - f_v \omega_m \tag{2.55}$$

where T_m is the mechanical torque from the turbine; J_m is the moment of inertia; f_v is the coefficient of friction and ω_m is the rotor mechanical speed.

According to the stator flux oriented vector control, the voltage vector coinciding with the q-axis is 90° ahead of the flux vector which coincides with d-axis when the stator resistance is neglected. Therefore, the flux and voltage equations of the stator can be simplified as:

$$\phi_{sd} = \phi_s$$

$$\phi_{sq} = 0 \tag{2.56}$$

where ϕ_s is the amplitude of stator flux vector. Then, the flux and the voltage of the rotor can be rewritten as:

$$v_{sd} = 0$$

$$v_{sq} = V_s = \omega_s \phi_{sd} \tag{2.57}$$

where V_s is the stator voltage amplitude. The parameters ϕ_s and V_s can be regarded as constant in the ideal grid. Consequently, the stator current can be rewritten as:

$$i_{sd} = \frac{\phi_s}{L_s} - \frac{L_m}{L_s} i_{rd}$$

$$i_{sq} = -\frac{L_m}{L_s} i_{rq} \tag{2.58}$$

The rotor voltage and flux should be developed as follows:

$$\phi_{rd} = \left(L_r - \frac{L_m^2}{L_s} \right) i_{rd} + \frac{L_m V_s}{L_s \omega_s}$$
$$\phi_{rq} = \left(L_r - \frac{L_m^2}{L_s} \right) i_{rq}$$

(2.59)

$$v_{rd} = R_r i_{rd} + \left(L_r - \frac{L_m^2}{L_s} \right) \frac{di_{rd}}{dt} - \omega_{sl} \left(L_r - \frac{L_m^2}{L_s} \right) i_{rq}$$
$$v_{rq} = R_r i_{rq} + \left(L_r - \frac{L_m^2}{L_s} \right) \frac{di_{rq}}{dt} + \omega_{sl} \left(L_r - \frac{L_m^2}{L_s} \right) i_{rd} + \frac{\omega_{sl}}{\omega_s} \frac{L_m V_s}{L_s}$$

(2.60)

where ω_{sl} is the slip electric angular velocity and $\omega_{sl} = \omega_s - \omega_r$.

2.2.5   Modeling of PMSG

The benefit of using a PMSG in TST is its higher efficiency compared to other conventional generators. Moreover, PMSG is rarely used in gear-driven systems which could further increase the availability of the system, reduce the active weight and the maintenance requirement. Although DFIG uses smaller power converters and has lower cost for large turbine application, the limited speed range would result lower energy yield than PMSG. A schematic diagram of PMSG-based TST is shown in Figure 2.18 [9,48].

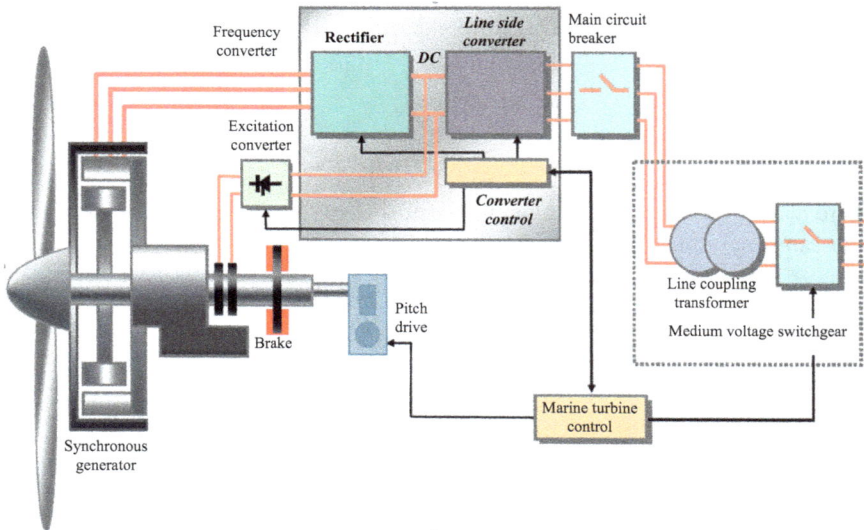

Figure 2.18   Schematic diagram of PMSG-based TST [9]

Indeed, the stator voltage equations in stationary reference frame are the same as DFIG in (2.49). The different part should be the flux equation. Thus, the flux equation for PMSG is given as follows:

$$[\phi_s]_3 = [L_{ss}][i_s]_3 + [\phi_{pm}]_3 \tag{2.61}$$

with

$$[L_{ss}] = \begin{bmatrix} L_0 + L_2\cos(2\theta_r) & M_0 + M_2\cos\left(2\left(\theta_r + \dfrac{2\pi}{3}\right)\right) \\ M_0 + M_2\cos\left(2\left(\theta_r + \dfrac{2\pi}{3}\right)\right) & L_0 + L_2\cos\left(2\left(\theta_r - \dfrac{2\pi}{3}\right)\right) \\ M_0 + M_2\cos\left(2\left(\theta_r - \dfrac{2\pi}{3}\right)\right) & M_0 + M_2\cos(2\theta_r) \end{bmatrix}$$

$$\begin{matrix} M_0 + M_2\cos\left(2\left(\theta_r - \dfrac{2\pi}{3}\right)\right) \\ M_0 + M_2\cos(2\theta_r) \\ L_0 + L_2\cos\left(2\left(\theta_r + \dfrac{2\pi}{3}\right)\right) \end{matrix}$$

$$[\phi_{pm}]_3 = \phi_1 \begin{bmatrix} \cos(\theta_r) \\ \cos\left(\theta_r - \dfrac{2\pi}{3}\right) \\ \cos\left(\theta_r + \dfrac{2\pi}{3}\right) \end{bmatrix}$$

where $[\phi_{pm}]_3$ is the permanent magnet (PM) flux linkage vector; L_0, M_0 are the constant part of the self and mutual inductances, respectively, normally $L_0 = -2M_0$; L_2, M_2 are the amplitude of the second harmonic of the self and mutual inductances separately, normally $L_2 = M_2$; if L_2 and M_2 are equal to zero, this machine should be surface PMSG, otherwise, it is salient-Pole PMSG.

According to Park transformation, the model of PMSG in the $d - q$ coordinate axis is summarized as:

$$\begin{aligned} v_{sd} &= R_s i_{sd} + \frac{d}{dt}\phi_{sd} - \omega_r\phi_{sq} \\ v_{sq} &= R_s i_{sq} + \frac{d}{dt}\phi_{sq} - \omega_r\phi_{sd} \end{aligned} \tag{2.62}$$

$$\begin{aligned} \phi_{sd} &= L_{sd}i_{sd} + \phi_1 \\ \phi_{sq} &= L_{sq}i_q \end{aligned} \tag{2.63}$$

where L_{sd}, L_{sq} are the components of stator inductance in the $d - q$ coordinate axis, $L_{sd} = L_0 - M_0 + (0.5L_1 + M_1)$, $L_{sq} = L_0 - M_0 - (0.5L_1 + M_1)$. For the PMSG, $\omega_s = \omega_r$. The electromagnetic torque of PMSG is given by the following formula:

$$T_{em} = \frac{3}{2}n_p \left(i_{sq}\phi_{sd} - \phi_{sq}i_{sd}\right)$$

$$= \frac{3}{2}n_p \left(\phi_1 i_{sq} + \left(L_{sd} - L_{sq}\right) i_{sd}i_{sq}\right) \tag{2.64}$$

For surface PMSG, this equation could be simplified as follows:

$$T_{em} = \frac{3}{2}n_p\phi_1 i_{sq} \tag{2.65}$$

2.2.5.1 Modeling of DSPMG

As conventional low speed, PMSG always has large machine diameter which may result in fouling in the complicated working condition. Other researchers also proposed some special low speed machines. One special low speed stator-PM machine, toothed pole DSPM machine, was proposed in [50]. This machine, seen in Figure 2.19, adopts teeth coupling technique which can decouple the operating frequency from the number of poles in the coil winding of the stator and allow for a notable growth in the frequency. Theoretically, the velocity of this kind of machine depends on the teeth of the rotor N_r and the pulsation ω_s as shown below:

$$\omega_m = \frac{\omega_s}{N_r} = \frac{2\pi f}{N_r} \tag{2.66}$$

where f is the electrical frequency. Meanwhile, this kind machine also has some other attractive characteristics: high torque density and power density; easy to control and cool; reduce the winding mass and the resistance. All these prominent advantages above make it quite applicable to TST, especially for the direct drive system [49]. The

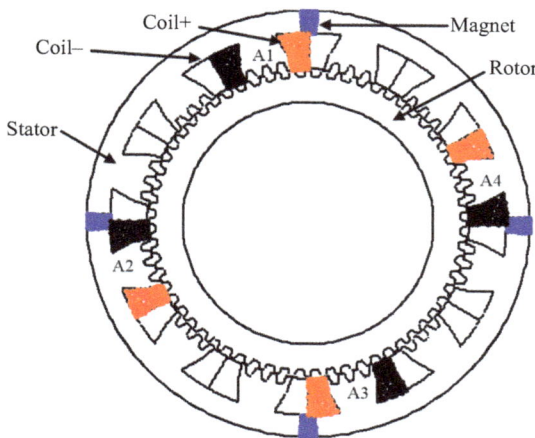

Figure 2.19   DSPMG structure [49]

basic stator flux equation is the same as (2.61). However, due to the special stator-PM structure, the inductance matrix and PM flux linkage vector are different from the conventional PMSG. They are defined as follows:

$$
[L_{ss}] = \frac{2}{3}
\begin{bmatrix}
L_0 + L_1 \cos(\theta_s) & M_0 + M_1 \cos\left(\theta_s + \frac{2\pi}{3}\right) \\
M_0 + M_1 \cos\left(\theta_s + \frac{2\pi}{3}\right) & L_0 + L_1 \cos\left(\theta_s - \frac{2\pi}{3}\right) \\
M_0 + M_1 \cos\left(\theta_s - \frac{2\pi}{3}\right) & M_0 + M_1 \cos(\theta_s)
\end{bmatrix}
$$

$$
\begin{matrix}
M_0 + M_1 \cos\left(\theta_s - \frac{2\pi}{3}\right) \\
M_0 + M_1 \cos(\theta_s) \\
L_0 + L_1 \cos\left(\theta_s + \frac{2\pi}{3}\right)
\end{matrix}
\tag{2.67}
$$

$$
[\phi_{pm}]_3 =
\begin{bmatrix}
\phi_0 + \phi_1 \cos(\theta_s) \\
\phi_0 + \phi_1 \cos\left(\theta_s - \frac{2\pi}{3}\right) \\
\phi_0 + \phi_1 \cos\left(\theta_s + \frac{2\pi}{3}\right)
\end{bmatrix}
\tag{2.68}
$$

where L_1 and M_1 are the fundamental of the self and mutual inductance; ϕ_0 is the constant part of the PM flux linkage. According to Park transformation, the flux of DSPMG in the $d - q$ coordinate axis and the voltage are given by:

$$
\begin{aligned}
\phi_{sd} &= L_{sd} i_{sd} + M_{dq} i_{sq} + \phi_1 \\
\phi_{sq} &= L_{sq} i_{sq} + M_{dq} i_{sd}
\end{aligned}
\tag{2.69}
$$

$$
\begin{aligned}
v_{sd} &= \left(R_s + 2\omega_s M_{dq}\right) i_{sd} - \omega_s \left(\frac{3L_{sd}}{2} - \frac{L_{sq}}{2}\right) i_{sq} + \\
& \quad L_{sd}\frac{di_{sd}}{dt} + M_{dq}\frac{di_{sq}}{dt} \\
v_{sq} &= \left(R_s - 2\omega_s M_{dq}\right) i_{sq} + \omega_s \left(\frac{3L_{sq}}{2} - \frac{L_{sd}}{2}\right) i_{sd} + \\
& \quad L_{sq}\frac{di_{sq}}{dt} + M_{dq}\frac{di_{sd}}{dt} + \phi_1 \omega_s
\end{aligned}
\tag{2.70}
$$

with

$$
\begin{aligned}
L_{sd} &= L_0 - M_0 + M_{dq} \cos(3\theta_s) \\
L_{sq} &= L_0 - M_0 - M_{dq} \cos(3\theta_s) \\
M_{dq} &= -\left(\frac{L_1}{2} + M_1\right)
\end{aligned}
\tag{2.71}
$$

where M_{dq} is the mutual inductance between d–q axes of DSPMG. The torque expression should be rewritten as:

$$T_{em} = \frac{3}{2}\left[N_r\phi_1 i_{sq} - \frac{N_r}{2}\left(L_{sd} - L_{sq}\right) i_{sd} i_{sq} + \frac{N_r}{2}M_{dq}\left(i_{sd}^2 - i_{sq}^2\right)\right] \quad (2.72)$$

2.3   Controller development for TST

2.3.1   HOSMC for DFIG-based TST

In order to obtain the maximum power from the tidal stream, a step-by-step procedure should be followed: first, the turbine speed reference ω_{ref} is generated by MPPT strategy [6]. Then, an optimal electromagnetic torque, which ensures the rotor speed convergence to ω_{m_ref}, is computed using the following mechanical equation:

$$T_{em\_ref} = T_m + f\omega - \alpha\left(\omega - \omega_{m\_ref}\right) + J\dot{\omega}_{m\_ref}, \quad (2.73)$$

where $\alpha > 0$ is a positive constant. Afterwards, rotor current references are derived to ensure that the DFIG torque and reactive power convergence to the optimal torque and zero, respectively:

$$
\begin{aligned}
i_{rq\_ref} &= \frac{-L_s}{n_p L_m}\frac{T_{em\_ref}}{\phi_{sd}}\\
i_{rd\_ref} &= \frac{\phi_{sd}}{L_m}
\end{aligned}
\quad (2.74)
$$

In order to ensure that the rotor currents convergence to their references, a robust high-order sliding mode strategy is used [12]. Let us define the following sliding mode surfaces:

$$
\begin{aligned}
s_1 &= i_{rd} - i_{rd\_ref}\\
s_2 &= i_{rq} - i_{rq\_ref}
\end{aligned}
\quad (2.75)
$$

It follows that

$$
\begin{aligned}
\dot{s}_1 &= \frac{L_s}{L_m^2 - L_r L_s}\left(v_{rd} + R_r i_{rd} - \omega_r\left(L_r i_{rq} + L_m i_{sq}\right) - \frac{L_m}{L_s}v_{sd} - \frac{L_m R_s}{L_s}i_{sd}\right)\\
&\quad + \frac{L_s}{L_m^2 - L_r L_s}\left(\frac{L_m}{L_s}\omega_s\left(L_s i_{sq} + L_m i_{rq}\right)\right) - \dot{i}_{rd\_ref}
\end{aligned}
\quad (2.76)
$$

$$\ddot{s}_1 = \varphi_1\left(t,x\right) + \gamma_1\left(t,x\right)\dot{v}_{rd}$$

$$
\begin{aligned}
\dot{s}_2 &= \frac{L_s}{L_m^2 - L_r L_s}\left(v_{rq} + R_r i_{rq} - \omega_r\left(L_r i_{rd} + L_m i_{sd}\right) - \frac{L_m}{L_s}v_{sq} - \frac{L_m R_s}{L_s}i_{sq}\right)\\
&\quad + \frac{L_s}{L_m^2 - L_r L_s}\left(\frac{L_m}{L_s}\omega_s\left(L_s i_{sd} + L_m i_{rd}\right)\right) - \dot{i}_{rq\_ref}
\end{aligned}
\quad (2.77)
$$

$$\ddot{s}_2 = \varphi_2\left(t,x\right) + \gamma_2\left(t,x\right)\dot{v}_{rq}$$

where $\phi_1(t,x)$, $\phi_2(t,x)$, $\gamma_1(t,x)$, and $\gamma_2(t,x)$ are uncertain functions which satisfy:

$$\begin{cases} \varphi_1 > 0, |\varphi_1| > \Phi_1, 0 < \Gamma_{m1} < \gamma_1 < \Gamma_{M1} \\ \varphi_2 > 0, |\varphi_2| > \Phi_2, 0 < \Gamma_{m2} < \gamma_2 < \Gamma_{M2} \end{cases}$$

The developed control approach which is based on the super twisting algorithm has been introduced by Levant [51]. The developed high-order (second) sliding mode controller contains two parts:

$$\begin{aligned} v_{rd} &= u_1 + u_2 \\ v_{rq} &= w_1 + w_2 \end{aligned} \tag{2.78}$$

In order to ensure the convergence of SM manifolds to zero in finite time, the gains can be chosen as follows [12,51]:

$$\begin{cases} \alpha_i > \dfrac{\Phi_i}{\Gamma_{mi}} \\ \beta_i^2 \geq \dfrac{4\Phi_i}{\Gamma_{mi}^2} \dfrac{\Gamma_{Mi}(\alpha_i + \Phi_i)}{\Gamma_{mi}(\alpha_i - \Phi_i)}, i = 1, 2 \\ 0 < \rho \leq 0.5 \end{cases} \tag{2.79}$$

The above developed HOSMC strategy for a DFIG-based TST is illustrated in Figure 2.20. With the filtered SHOM model resource (see in Figure 2.21), the DFIG-based TST control performances are shown in Figures 2.22–2.24, respectively, illustrating the rotor speed tracking performance, the reactive, and the generated active power. The obtained results show good tracking performances.

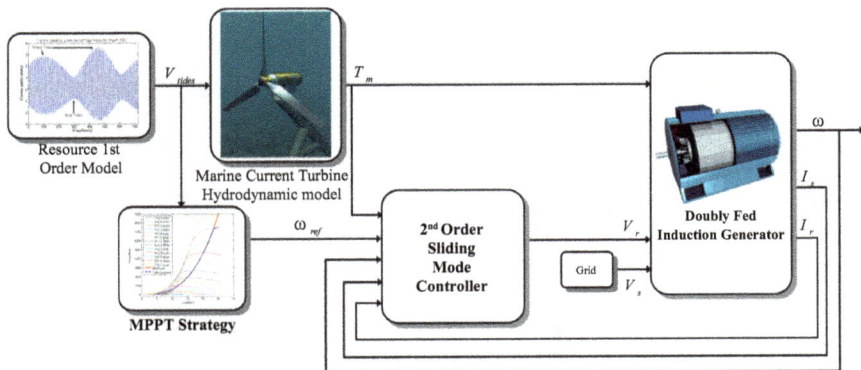

Figure 2.20 The developed control structure of DFIG-based TST [6]

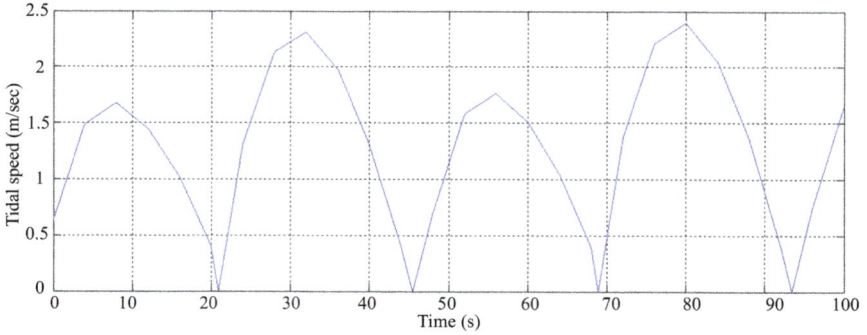

Figure 2.21   Filtered resource speed based on SHOM [6]

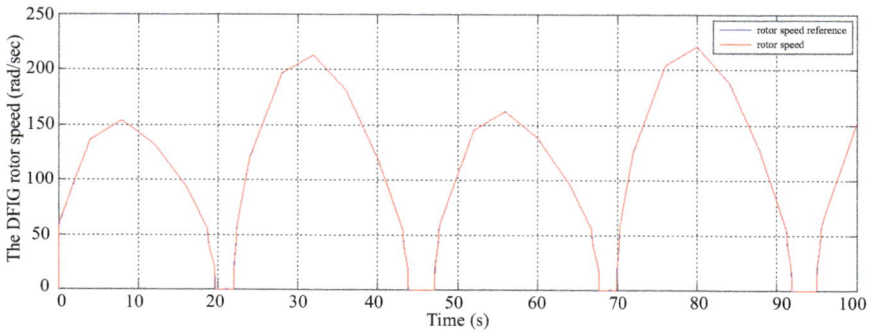

Figure 2.22   The DFIG rotor speed and its reference [6]

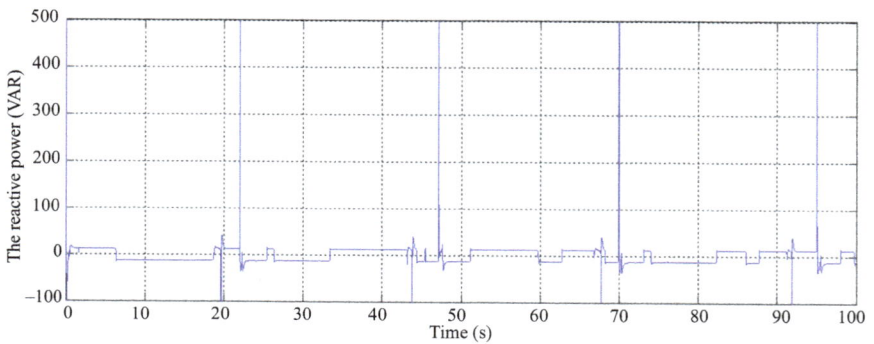

Figure 2.23   The DFIG reactive power [6]

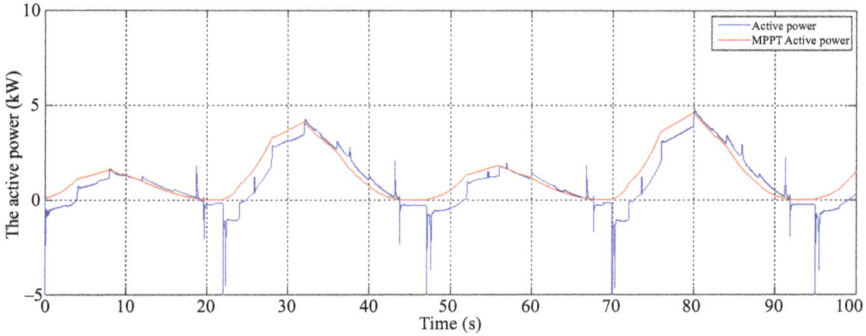

Figure 2.24   The DFIG active power [6]

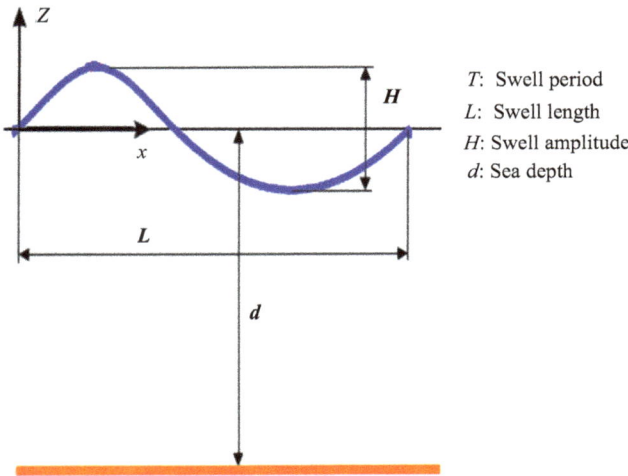

Figure 2.25   Swell characteristics [6]

Since the swell effect can significantly perturb the simplified first-order resource model, a swell stoke model is added. In this case, the speed potential is given by (2.80) (see in Figure 2.25) [40]

$$V_{swell} = \frac{HL}{2T} \frac{\cosh\left(2\pi\left(\frac{z+d}{L}\right)\right)}{\sinh\left(2\pi\left(dL\right)\right)} \sin\left(2\pi\left(\frac{t}{T} - \frac{x}{L}\right)\right) \tag{2.80}$$

The turbulent resource characteristics are then given in Figure 2.26. In this case, the DFIG-based TST control performances are shown in Figures 2.27–2.29, respectively, illustrating the rotor speed tracking performance, the reactive, and the generated active power. In spite of the good tracking accuracy of HOSMC, the generated power

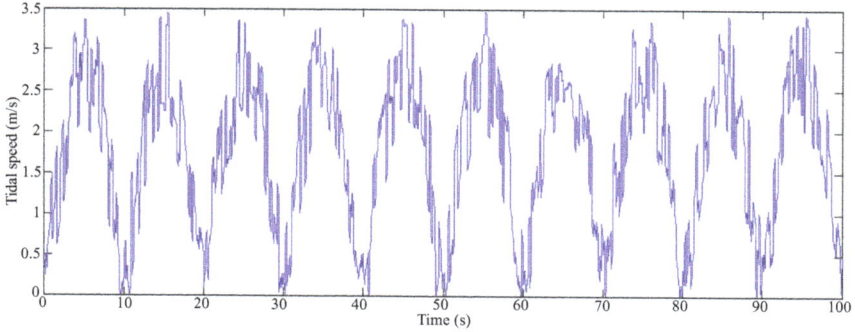

Figure 2.26   Turbulent resource speed [6]

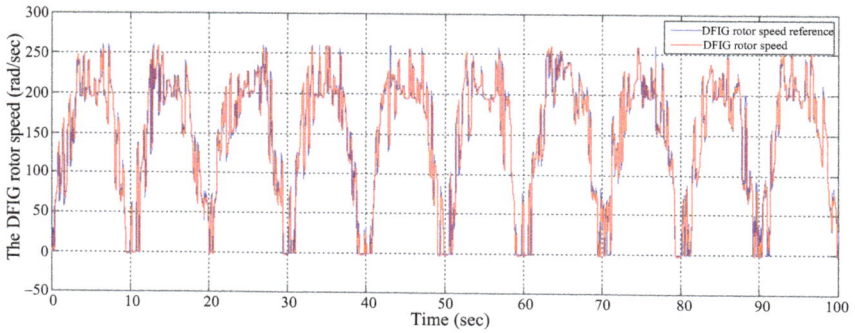

Figure 2.27   The DFIG rotor speed and its reference [6]

exhibits fluctuations. In fact, these fluctuations are mainly due to accelerations and decelerations imposed to the tidal turbine by the MPPT strategy for a turbulent resource. This problem could be simply solved by filtering the resource at a certain level to avoid losing useful powers.

2.3.2   HOSMC for PMSG-based TST

The optimal electromagnetic torque of PMSG-based TST is similar to that of DFIG-based TST as in (2.73), the only difference is the current references that ensure the PMSG torque convergence to the optimal torque. Consequently, the current references of PMSG are summarized as follows:

$$i_{sd\_ref} = 0$$

$$i_{sq\_ref} = \frac{2}{3} \frac{T_{em}}{n_p \phi_1}$$

(2.81)

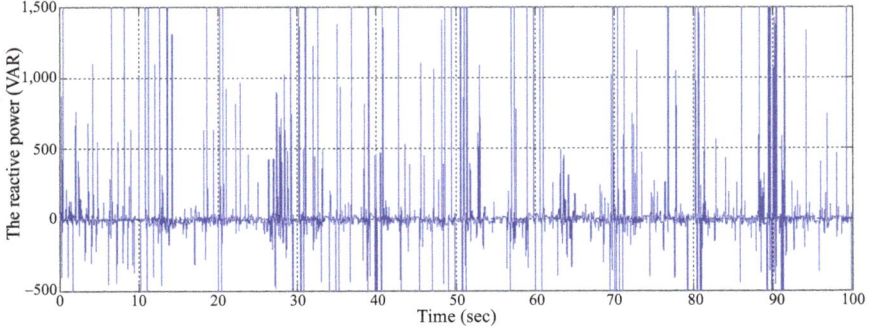

Figure 2.28   The DFIG reactive power [6]

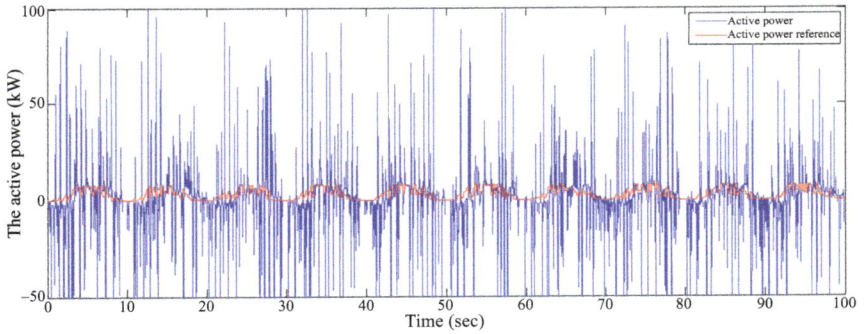

Figure 2.29   The DFIG active power [6]

In order to ensure that the currents convergence to their references, HOSMC strategy is also used. The following sliding surfaces are defined which are similar to (2.75):

$$s_1 = i_{sd} - i_{sd\_ref}$$
$$s_2 = i_{sq} - i_{sq\_ref}$$

(2.82)

It follows that

$$\dot{s}_1 = \dot{i}_{sd} - \dot{i}_{sd\_ref}$$
$$\ddot{s}_1 = \varphi_1(t,x) + \gamma_1(t,x)\dot{v}_{sd}$$
and

(2.83)

$$\dot{s}_2 = \dot{i}_{sq} - \dot{i}_{sq\_ref}$$
$$\ddot{s}_2 = \varphi_2(t,x) + \gamma_2(t,x)\dot{v}_{sq}$$

(2.84)

where $\varphi_1\,(t,x)$, $\varphi_2\,(t,x)$, $\gamma_1\,(t,x)$, and $\gamma_2\,(t,x)$ are uncertain bounded functions that satisfy:

$$\begin{cases} \varphi_1 > 0, & |\varphi_1| > \Phi_1, 0 < \Gamma_{m1} < \gamma_1 < \Gamma_{M1} \\ \varphi_2 > 0, & |\varphi_2| > \Phi_2, 0 < \Gamma_{m2} < \gamma_2 < \Gamma_{M2} \end{cases}$$

This developed SOSMC also contains two parts:

$$\begin{aligned} v_{sd} &= u_1 + u_2 \\ v_{sq} &= w_1 + w_2 \end{aligned} \tag{2.85}$$

with

$$\begin{cases} \dot{u}_1 &= -\alpha_1 \text{sign}\,(s_1) \\ u_2 &= -\beta_1\,|s_1|^\rho\,\text{sign}\,(s_1) \\ \dot{w}_1 &= -\alpha_2 \text{sign}\,(s_2) \\ w_2 &= -\beta_2\,|s_2|^\rho\,\text{sign}\,(s_2) \end{cases}$$

Again, in order to ensure that the convergence of SM manifolds to zero in finite time, the gains can be chosen as follows, which are the same as DFIG and given by:

$$\begin{cases} \alpha_i > \dfrac{\Phi_i}{\Gamma_{mi}} \\ \beta_i^2 > \dfrac{4\Phi_i}{\Gamma_{mi}^2}\dfrac{\Gamma_{mi}\,(\alpha_i + \Phi_i)}{\Gamma_{Mi}\,(\alpha_i - \Phi_i)}, i = 1, 2 \\ 0 < \rho \leq 0.5 \end{cases} \tag{2.86}$$

The above developed second-order sliding mode control strategy for a PMSG-based marine current turbine is illustrated by the block diagram in Figure 2.30. In this case, the tidal current is also simulated as a first-order model as in Figure 2.21. Therefore, the PMSG-based TST control performances are shown in Figures 2.31–2.33, respectively, illustrating the current, the rotor speed, and the generated power. The obtained results show good tracking performances of the PMSG current and rotor speed.

2.3.3   HOSMC for DSPMG-based TST

First, DSPMG could be regarded as a controlled nonlinear system, whose state equation is given by:

$$\begin{aligned} \dot{x} &= f\,(x, t) + g\,(x, t)\,u \\ y &= s\,(x, t) \end{aligned} \tag{2.87}$$

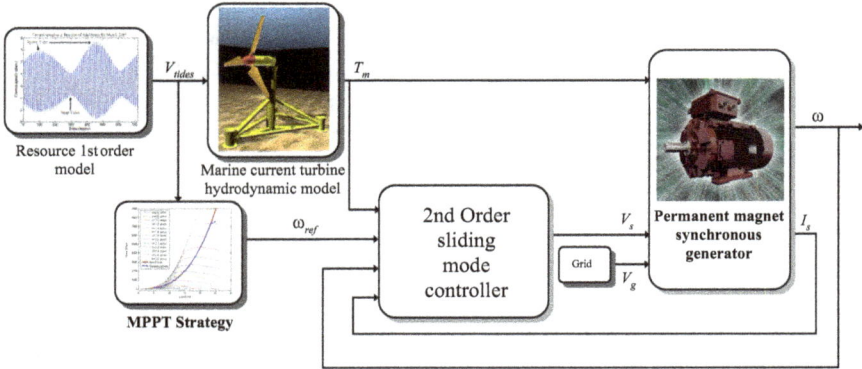

Figure 2.30   The developed control structure of PMSG-based TST [9]

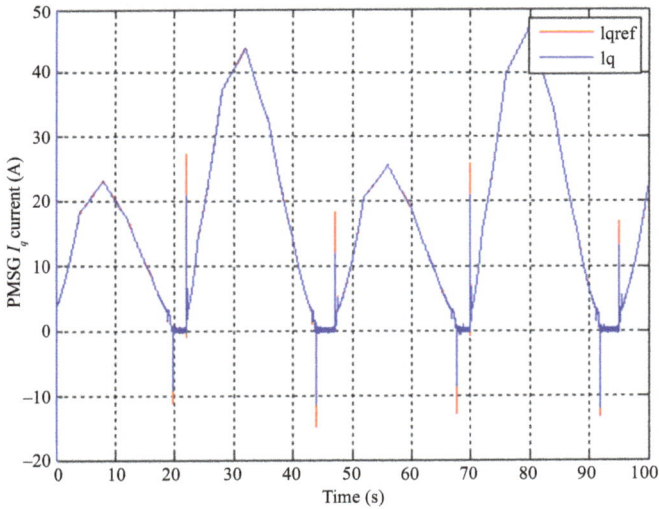

Figure 2.31   The PMSG i_{sq} current tracking performances [9]

where x is the system state variables; u is the control input; $f(x, t)$ and $g(x, t)$ are both smooth uncertain continuous functions, respectively. Then, the system variables are defined as:

$$x_1 = x_{ref} - x_{actual}$$

$$x_2 = \int_{-\infty}^{t} x_1 dt = \int_{-\infty}^{t} (x_{ref} - x_{actual}) \, dt$$

(2.88)

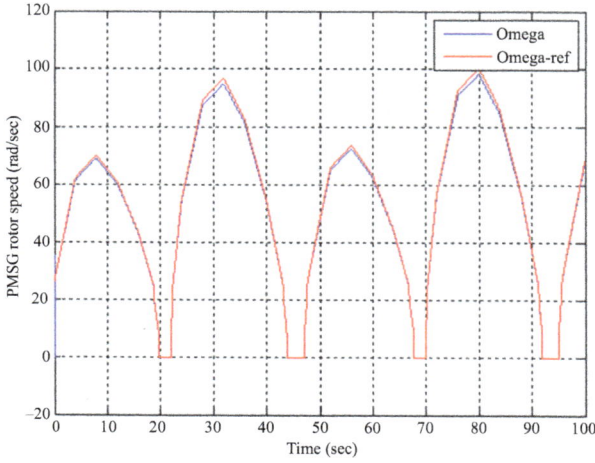

Figure 2.32   The PMSG rotor speed tracking performances [9]

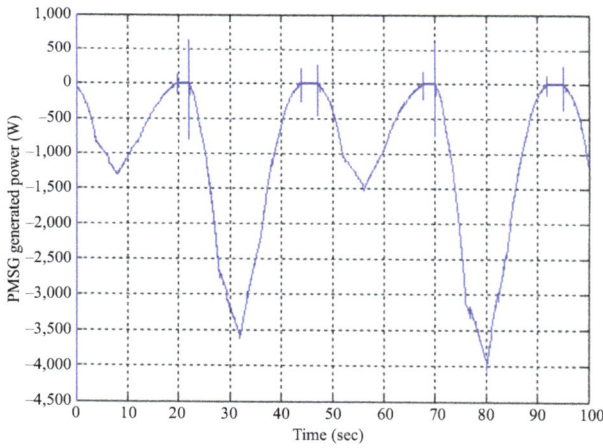

Figure 2.33   The PMSG generated power [9]

with

$$
\begin{cases}
x_{ref} &= \left[\ \omega_{ref}\ i_{sd\_ref}\ i_{sq\_ref}\ \right]^T \\
x_{actual} &= \left[\ \omega_{ref}\ i_{sd\_ref}\ i_{sq\_ref}\ \right]^T \\
x_1 &= \left[\ x_{1\omega_m}\ x_{1i_{sd}}\ x_{1i_{sq}}\ \right]^T \\
&= \left[\ \omega_{m\_ref} - \omega_m\ i_{sd\_ref} - i_{sd}\ i_{sq\_ref} - i_{sq}\ \right]^T
\end{cases}
$$

where x_{ref} and x_{actual} are the theoretical reference value and the actual measurement. In this section, ISMS is designed under the utilization of system variables x_1 and x_2 as shown below:

$$s = x_1 + k_i \int_{-\infty}^{t} x_1 dt = x_1 + k_i x_2 \tag{2.89}$$

Then, the time-derivative of the sliding mode surface in (2.89) can be obtained as shown below in combination with (2.70):

$$
\begin{aligned}
\dot{s}_{\omega m} &= \dot{\omega}_{m\_ref} - \dot{\omega}_m + k_{i\omega m} x_{1\omega m} \\
&= \dot{\omega}_{m\_ref} - \frac{1}{J_m}[T_{em} - T_m - f_v \omega_m] + k_{i\omega m} x_{1\omega m} \\
\dot{s}_{i_{sd}} &= \dot{i}_{sd\_ref} - \dot{i}_{sd} + k_{ii_d} x_{1i_d} \\
&= \dot{i}_{sd\_ref} - \frac{1}{L_{sd}}\left[v_{sd} - (R_s + 2\omega_s M_{dq}) i_{sd} - M_{dq}\frac{di_{sq}}{dt}\right] + \\
&\quad \frac{1}{L_{sd}}\left[\omega_s\left(\frac{3}{2}L_{sd} - \frac{1}{2}L_{sq}\right)i_{sq}\right] + k_{ii_d} x_{1i_{sd}} \\
\dot{s}_{i_{sq}} &= \dot{i}_{sq\_ref} - \dot{i}_{sq} + k_{ii_q} x_{1i_q} \\
&= \dot{i}_{sq\_ref} - \frac{1}{L_{sq}}\left[v_{sq} - (R_s - 2\omega_s M_{dq}) i_{sq} - M_{dq}\frac{di_{sd}}{dt}\right] + \\
&\quad \frac{1}{L_{sq}}\left[\omega_s\left(\frac{3}{2}L_{sq} - \frac{1}{2}L_{sd}\right)i_{sd} - \phi_1 \omega_s\right] + k_{ii_q} x_{1i_{sq}}
\end{aligned}
\tag{2.90}
$$

where $S_{\omega m}$, $S_{i_{sd}}$, and $S_{i_{sq}}$ are ISMSs of the speed and current controllers, respectively; $x_{1\omega m}$, x_{1i_d}, and x_{1i_q} are the tracking errors of the speed and current controllers, respectively. In order to make the system have global robustness, the system should be on ISMS at the initial time of the system ($t = 0$). Thus, an integral initial value can be set as follows:

$$I_0 = \int_{-\infty}^{0} x_1(\tau)\, dt = -\frac{x_1(0)}{k_i} \tag{2.91}$$

where x_0 is the initial value of x_1. For SOSMC, the control objective of the system is to make the system state reach SMS ($s = 0$) in a finite time and have SOSM ($s = \dot{s} = 0$). Considering the nonlinearity and uncertainty of the system, the following controllable first-order system is considered to replace the error tracking system of the speed and the current in the synchronous reference $d - q$ frame as given below:

$$\dot{x}_1 = u(t) + \xi(t) \tag{2.92}$$

where $\xi(t)$ is the uncertain disturbance term of the system; $u(t)$ is the super-twisting control law which is given in (2.93). Combining (2.21), (2.39), (2.90), and (2.92), the output of the speed and current controllers should be as follows [38]:

$$T_{em} = J_m \left[\int \alpha_{\omega m} \mathrm{sat} \left(s_{\omega m} \right) dt + \beta_{\omega m} \left| s_{\omega m} \right|^{\frac{1}{2}} + \dot{\omega}_{ref} - \xi_{\omega m}(t) \right] +$$

$$f_v \omega_m + T_m$$

$$v_{sd} = L_{sd} \left[\int \alpha_{i_{sd}} \mathrm{sat} \left(s_{i_{sd}} \right) dt + \beta_{i_{sd}} \left| s_{i_{sd}} \right|^{\frac{1}{2}} \mathrm{sat} \left(s_{i_{sd}} \right) + \dot{i}_{sd\_ref} - \xi_{i_{sd}}(t) \right] +$$

$$\left(R_s + 2\omega_s M_{dq} \right) i_{sd} + M_{dq} \frac{d i_{sq}}{dt} - \omega_s \left(1.5 L_{sd} - 0.5 L_{sq} \right) i_{sq}$$

(2.93)

$$v_{sq} = L_{sq} \left[\int \alpha_{i_{sq}} \mathrm{sat} \left(s_{i_{sq}} \right) dt + \beta_{i_{sq}} \left| s_{i_{sq}} \right|^{\frac{1}{2}} \mathrm{sat} \left(s_{i_{sq}} \right) + \dot{i}_{sq\_ref} - \xi_{i_{sq}}(t) \right] +$$

$$\left(R_s - 2\omega_s M_{dq} \right) i_{sq} + M_{dq} \frac{d i_{sd}}{dt} + \omega_s \left(1.5 L_{sd} - 0.5 L_{sq} \right) i_{sq} + \phi_1 \omega_s$$

It is to be noted here that saturation function is used in (2.93) to further reduce the chattering effect. As this DSPMG is nonconventional, the current waveform should also be mentioned. In order to obtain constant torque, some current harmonics should be injected. If the quasi-sinusoidal current in a three-phase stationary coordinate system is as shown below:

$$i_a = -I_m (\theta_e) \sin (\theta_e + \theta_0)$$

$$i_b = -I_m (\theta_e) \sin \left(\theta_e - \frac{2\pi}{3} + \theta_0 \right)$$

(2.94)

$$i_b = -I_m (\theta_e) \sin \left(\theta_e + \frac{2\pi}{3} + \theta_0 \right)$$

where $I_m (\theta_e)$ is the quasi-sinusoidal current amplitude, which varies with electrical angle θ_e. Then, the variable current amplitude, which could produce the constant torque, can be developed as [52]:

$$I_m (\theta_e) = \frac{-\frac{3}{2} N_r \phi_1 \cos (\theta_0)}{\frac{3}{2} N_r \left(\frac{L_1}{2} + M_1 \right) \sin (3\theta_e + 2\theta_0)} +$$

$$\frac{\sqrt{\left(\frac{3}{2} N_r \varphi_1 \cos (\theta_0) \right)^2 + 3 N_r \left(\frac{L_1}{2} + M_1 \right) \sin (3\theta_e + 2\theta_0) T_{em}}}{\frac{3}{2} N_r \left(\frac{L_1}{2} + M_1 \right) \sin (3\theta_e + 2\theta_0)}$$

(2.95)

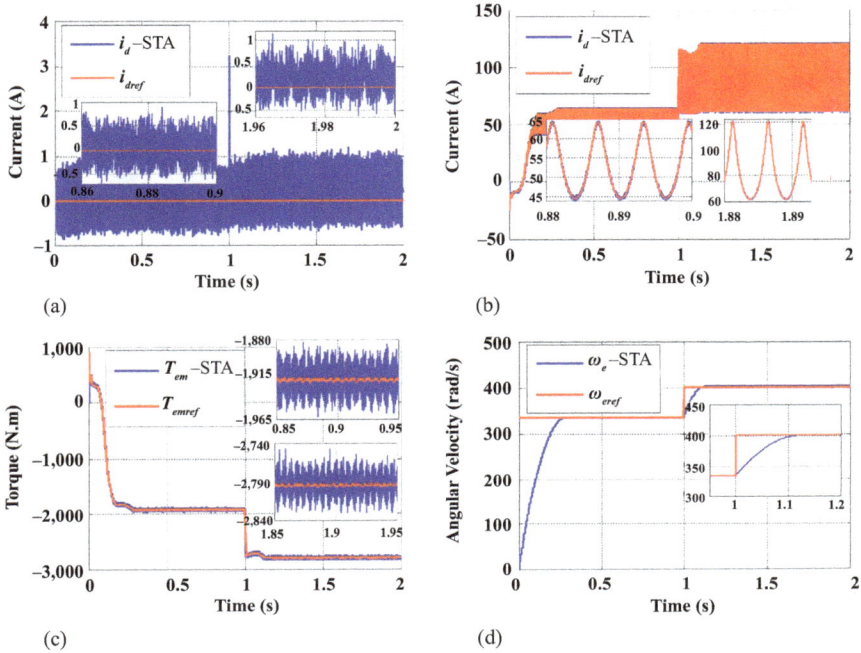

Figure 2.34 Verification of STA-based HOSMC [38]. (a) d-axis current. (b) q-axis current. (c) Electromagnetic torque. (d) Electrical angular velocity.

In order to verify the accuracy and robustness of SOSMC as given in (2.95), simulation studies are carried out and the parameters are given in [38]. In this simulation, the given tidal current velocity is 2.5 m/s at the beginning; then, it is abruptly changed to 3.0 m/s at $t = 1$ s and remains at this value until the end of simulation. From Figure 2.34(a) and (b), although the model of DSPMG is nonlinear, SOSMC still has very good control performance. The tracking errors of i_{sd} and i_{sq} are very small, even when the speed increases to 3.0 m/s. Due to the complicated relationship between the current and torque, even a small current error will cause large electromagnetic torque fluctuation. From Figure 2.34(c), the error of electromagnetic torque can be controlled within 2.0% and even smaller in the second stage. Indeed, the torque fluctuation would bring speed fluctuation. From Figure 2.34(d), the speed error is far less than 1%.

In this part, the robustness of SOSMC will be discussed. The values of R_s and L in the machine model will be changed from 50% to 200% of the initial value. The electromagnetic torque and electrical angular velocity are selected and shown in Figures 2.35 and 2.36, respectively. From Figure 2.35(a) and (b), STA-based SOSMC can still approach the reference at the same response time and track the reference well with the great variation of the resistance R_s. Evidently, the changes in both electrical angular velocity and electromagnetic torque caused by the variation in resistance are

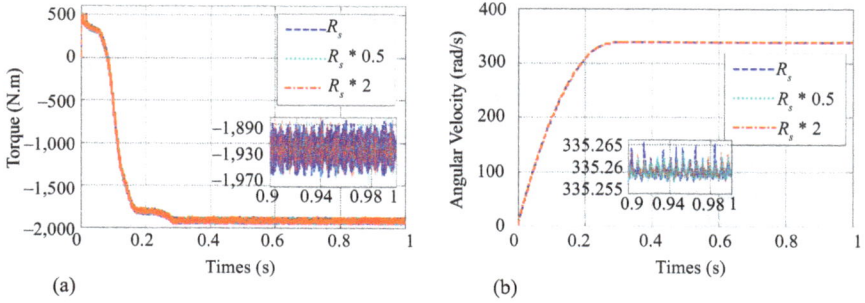

Figure 2.35   *Influence of R_s [38]. (a) Electromagnetic torque. (b) Electrical angular velocity.*

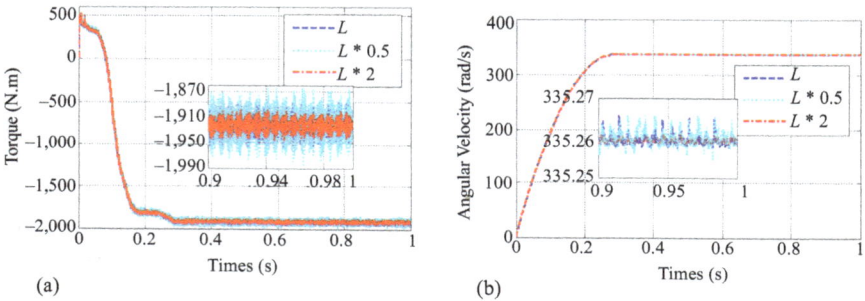

Figure 2.36   *Influence of L (L_0 and L_1) [38]. (a) Electromagnetic torque. (b) Electrical angular velocity.*

negligible. From Figure 2.36(a) and (b), it is found that when the inductance reduce, the errors would be smaller, otherwise, it will be bigger. The change of inductance has little effect on the control effect. According to these simulation results, they can well validate the robustness to the significant variation of parameters under the STA-based SOSMC.

2.4   Conclusion

Tidal energy sources can provide predictable and stable power output. This has made this technology as an attractive solution to mitigate the variable electricity production from wind and solar. However, as this technology operates in sea, tidal stream turbines face the harsh marine environment condition. This complicates the control of tidal stream turbine. This chapter presented an overview of tidal stream turbine control system. First, an overview of the feedback control as applied to the TSTs is presented. Two popular linear and nonlinear control methods are analyzed. Then, the modeling part is presented, where tidal resources, tidal turbine, and tidal electricity generation system, e.g., generator are presented with appropriate mathematical equations. Then, sliding mode controllers are designed for three generator topology. Results are also provided to demonstrate the performance of sliding mode controllers. Results on various generator topologies show that SMC is a very suitable choice for TSTs. This control method can provide fast and accurate tracking, comes with very good robustness to system parameter variations and has excellent external disturbance rejection property.

References

[1] Ben Elghali S, Benbouzid MEH, and Charpentier JF. Generator systems for marine current turbine applications: a comparative study. *IEEE Journal of Oceanic Engineering*. 2012;37(3):554–563.

[2] Chen H, Xie W, Chen X, *et al.* Fractional-order PI control of DFIG-based tidal stream turbine. *Journal of Marine Science and Engineering*. 2020;8(5):309.

[3] Ghefiri K, Garrido AJ, Rusu E, *et al.* Fuzzy supervision based-pitch angle control of a tidal stream generator for a disturbed tidal input. *Energies*. 2018;11(11):2989.

[4] Gaamouche R, Redouane A, Belhorma B, *et al.* Optimal feedback control of nonlinear variable-speed marine current turbine using a two-mass model. *Journal of Marine Science and Application*. 2020;19(1):83–95.

[5] Zhou Z, Ben Elghali S, Benbouzid M, *et al.* Control strategies for tidal stream turbine systems—a comparative study of ADRC, PI, and high-order sliding mode controls. In: *IECON 2019—45th Annual Conference of the IEEE Industrial Electronics Society*, vol. 1. IEEE; 2019. p. 6981–6986.

[6] Ben Elghali S, Benbouzid MEH, Ahmed-Ali T, *et al.* High-order sliding mode control of a marine current turbine driven doubly-fed induction generator. *IEEE Journal of Oceanic Engineering*. 2010;35(2):402–411.

[7] Zhou Z, Ben Elghali S, Benbouzid M, *et al.* Tidal stream turbine control: an active disturbance rejection control approach. *Ocean Engineering*. 2020;202:107190.

[8] Ben Elghali S, Benbouzid MEH, and Charpentier JF. Modelling and control of a marine current turbine-driven doubly fed induction generator. *IET Renewable Power Generation*. 2010;4(1):1–11.

[9] Ben Elghali S, Benbouzid MEH, Charpentier JF, *et al.* High-order sliding mode control of a marine current turbine driven permanent magnet synchronous generator. In: *2009 IEEE International Electric Machines and Drives Conference*; 2009. p. 1541–1546.

[10] Ahmed H, Rios H, and Benbouzid M. Continuous sliding-mode control of tidal stream turbine. In: *14th IFAC Conference on Control Applications in Marine Systems, Robotics, and Vehicles*; 2022.

[11] Chen H, Wang X, Benbouzid M, *et al.* Improved fractional-order PID controller of a PMSM-based wave compensation system for offshore ship cranes. *Journal of Marine Science and Engineering.* 2022;10(9):1238.

[12] Shtessel Y, Edwards C, Fridman L, *et al. Sliding Mode Control and Observation*, vol. 10. Springer, New York, NY; 2014.

[13] Slotine JJE and Li W. *Applied Nonlinear Control*, vol. 199. Prentice Hall, Englewood Cliffs, NJ; 1991.

[14] Chern TL and Wu YC. Design of integral variable structure controller and application to electrohydraulic velocity servosystems. In: *IEE Proceedings D (Control Theory and Applications)*, vol. 138. IET; 1991. p. 439–444.

[15] Castanos F and Fridman L. Analysis and design of integral sliding manifolds for systems with unmatched perturbations. *IEEE Transactions on Automatic Control.* 2006;51(5):853–858.

[16] Hung CP. Integral variable structure control of nonlinear system using a CMAC neural network learning approach. *IEEE Transactions on Systems, Man, and Cybernetics, Part B (Cybernetics).* 2004;34(1):702–709.

[17] Lee JH. Highly robust position control of BLDDSM using an improved integral variable structure systems. *Automatica.* 2006;42(6):929–935.

[18] Li P and Zheng ZQ. Sliding mode control approach with nonlinear integrator. *Control Theory & Applications.* 2011;28(3):421–426.

[19] Stepanenko Y, Cao Y, and Su CY. Variable structure control of robotic manipulator with PID sliding surfaces. *International Journal of Robust and Nonlinear Control.* 1998;8(1):79–90.

[20] Zak M. Terminal attractors in neural networks. *Neural Networks.* 1989;2(4):259–274.

[21] Venkataraman S and Gulati S. Terminal sliding modes: a new approach to nonlinear control synthesis. In: *Fifth International Conference on Advanced Robotics' Robots in Unstructured Environments.* IEEE, New York, NY; 1991. p. 443–448.

[22] Venkataraman S and Gulati S. Terminal slider control of robot systems. *Journal of Intelligent and Robotic Systems.* 1993;7(1):31–55.

[23] Venkataraman S and Gulati S. Control of nonlinear systems using terminal sliding modes. *Journal of Systems Engineering and Electronics.* 1993;29: 571–579.

[24] Yu X, Zhihong M, and Wu Y. Terminal sliding modes with fast transient performance. In: *Proceedings of the 36th IEEE Conference on Decision and Control*, vol. 2. IEEE, New York, NY; 1997. p. 962–963.

[25] Gao W and Hung JC. Variable structure control of nonlinear systems: a new approach. *IEEE Transactions on Industrial Electronics.* 1993;40(1):45–55.

[26] Slotine JJ and Sastry SS. Tracking control of non-linear systems using sliding surfaces, with application to robot manipulators. *International Journal of Control.* 1983;38(2):465–492.

[27] Fridman L and Levant A. Higher order sliding modes as a natural phenomenon in control theory. In: *Robust Control via Variable Structure and Lyapunov Techniques.* Springer, New York, NY; 1996. p. 107–133.

[28] Levant A. *Higher Order Sliding Modes and Their Application for Controlling Uncertain Processes*. Institute for System Studies of the USSR Academy of Science, Moskau, Dissertation, 1987.

[29] Emelyanov SV, Korovin SK, and Levantovsky LV. Second order sliding modes in controlling uncertain systems. *Soviet Journal of Computer and System Science*. 1986;24(4):63–68.

[30] Levant A. Sliding order and sliding accuracy in sliding mode control. *International Journal of Control*. 1993;58(6):1247–1263.

[31] Khalil HK. *Nonlinear Control*, vol. 406. Pearson, New York, NY; 2015.

[32] Levant A. Principles of 2-sliding mode design. *Automatica*. 2007;43(4): 576–586.

[33] Bartolini G, Ferrara A, and Usai E. Chattering avoidance by second-order sliding mode control. *IEEE Transactions on Automatic Control*. 1998;43(2): 241–246.

[34] Bartolini G, Pisano A, Punta E, *et al*. A survey of applications of second-order sliding mode control to mechanical systems. *International Journal of Control*. 2003;76(9-10):875–892.

[35] Fraenkel PL. Power from marine currents. *Proceedings of the Institution of Mechanical Engineers, Part A: Journal of Power and Energy*. 2002;216(1): 1–14.

[36] Love S. *A Channel Model Approach to Determine Power Supply Profiles and the Potential for Embedded Generation*. Ph.D. Dissertation, 2005.

[37] Yue F, Byung Ho C, and Guo-hong F. Global ocean tides from Geosat altimetry by quasi-harmonic analysis. *Chinese Journal of Oceanology and Limnology*. 2000;18(3):193–198.

[38] Chen H, Tang S, Han J, *et al*. High-order sliding mode control of a doubly salient permanent magnet machine driving marine current turbine. *Journal of Ocean Engineering and Science*. 2021;6(1):12–20.

[39] Ben Elghali S, Benbouzid MEH, Charpentier JF, *et al*. Experimental validation of a marine current turbine simulator: application to a permanent magnet synchronous generator-based system second-order sliding mode control. *IEEE Transactions on Industrial Electronics*. 2011;58(1):118–126.

[40] Batten W, Bahaj A, Molland A, *et al*. The prediction of the hydrodynamic performance of marine current turbines. *Renewable Energy*. 2008;33(5): 1085–1096.

[41] Bahaj A, Molland A, Chaplin J, *et al*. Power and thrust measurements of marine current turbines under various hydrodynamic flow conditions in a cavitation tunnel and a towing tank. *Renewable Energy*. 2007;32(3): 407–426.

[42] Ben Elghali S, Balme R, Le Saux K, *et al*. A simulation model for the evaluation of the electrical power potential harnessed by a marine current turbine. *IEEE Journal of Oceanic Engineering*. 2007;32(4): 786–797.

[43] Chen H, Tang T, Aït-Ahmed N, *et al*. Attraction, challenge and current status of marine current energy. *IEEE Access*. 2018;6:12665–12685.

[44]　Djebarri S, Charpentier JF, Scuiller F, *et al*. Design and performance analysis of double stator axial flux PM generator for Rim driven marine current turbines. *IEEE Journal of Oceanic Engineering*. 2016;41(1):50–66.

[45]　Chen H, Tang T, Aït-Ahmed N, *et al*. Generators for marine current energy conversion system: a state of the art review. In: *IECON 2017 – 43rd Annual Conference of the IEEE Industrial Electronics Society*; 2017. p. 2504–2509.

[46]　Djebarri S, Charpentier JF, Scuiller F, *et al*. Rough design of a double-stator axial flux permanent magnet generator for a rim-driven marine current turbine. In: *2012 IEEE International Symposium on Industrial Electronics*; 2012. p. 1450–1455.

[47]　Zhang J, Moreau L, Guo J, *et al*. Electromagnetic structure limits and control of a double stator permanent magnet generator for tidal energy applications. In: *IEEE International Power Electronics and Application Conference and Exposition*; 2014.

[48]　Zhou Z, Scuiller F, Charpentier JF, *et al*. Power smoothing control in a grid-connected marine current turbine system for compensating swell effect. *IEEE Transactions on Sustainable Energy*. 2013;4(3):816–826.

[49]　Chen H, Tang T, Han J, *et al*. Current waveforms analysis of toothed pole Doubly Salient Permanent Magnet (DSPM) machine for marine tidal current applications. *International Journal of Electrical Power & Energy Systems*. 2019;106:242–253.

[50]　Chen H, At-Ahmed N, Machmoum M, *et al*. Modeling and vector control of marine current energy conversion system based on doubly salient permanent magnet generator. *IEEE Transactions on Sustainable Energy*. 2016;7(1):409–418.

[51]　Levant A. Higher-order sliding modes, differentiation and output-feedback control. *International Journal of Control*. 2003;76(9-10):924–941.

[52]　Chen H, Li Q, Tang S, *et al*. Adaptive super-twisting control of doubly salient permanent magnet generator for tidal stream turbine. *International Journal of Electrical Power & Energy Systems*. 2021;128:106772.

Chapter 3
Tidal stream turbine fault–tolerant control

Elhoussin Elbouchikhi[1], Yassine Amirat[1], Gang Yao[2] and Mohamed Benbouzid[3]

3.1 Introduction

Tidal stream energy is highly predictable renewable energy resource and is considered as a promising technology to lower greenhouse gas emissions for sustainable development [1]. Unfortunately, tidal stream turbines are still in development stages with few industrial project deployments. Indeed, tidal energy systems are not yet massively deployed as they are facing reliability and availability challenges mainly due the harsh immersion conditions [2]. Even though, technological similarities exist between wind and tidal turbines, tidal turbines are not as mature as wind turbines technology and several design and operation differences require more investigation such as biofouling and marine current turbulence [3]. Moreover, tidal turbines are submerged systems, which have to withstand high loading and harsh submarine conditions that make this technology suffering from higher failures rate. In fact, load variability and rotor speed variations in tidal turbines lead to higher failure rate of major tidal turbines components as shown in Figure 3.1 [4]. Evidences of these failures highlight the need for specific condition monitoring scheme. Moreover, the control algorithm and energy management systems should be upgraded in order to be fault–tolerant and ensure safe operation under faulty conditions.

The adopted drivetrain and generator options affect the availability and consequently energy cost [5]. In particular, the direct-drive option is typically adopted to minimize maintenance issues by removing the gearbox and using an unconventional low speed and high diameter permanent magnet synchronous generator [6,7]. This unconventional generator type, apart from being subject to the same type of failures as conventional machines, can be subjected to magnet failures as it typically uses a relatively high number of permanent magnets [8,9]. Indeed, their magnetic force can weaken locally or uniformly [10]. Moreover, maximum power point tracking (MPPT) strategy is required to maximize the energy conversion process. To do so, flow-meters

[1]ISEN Yncréa Ouest, L@bISEN, France
[2]Shanghai Maritime University, Logistics Engineering College, China
[3]University of Brest, CNRS, Institut de Recherche Dupuy de Lôme, France

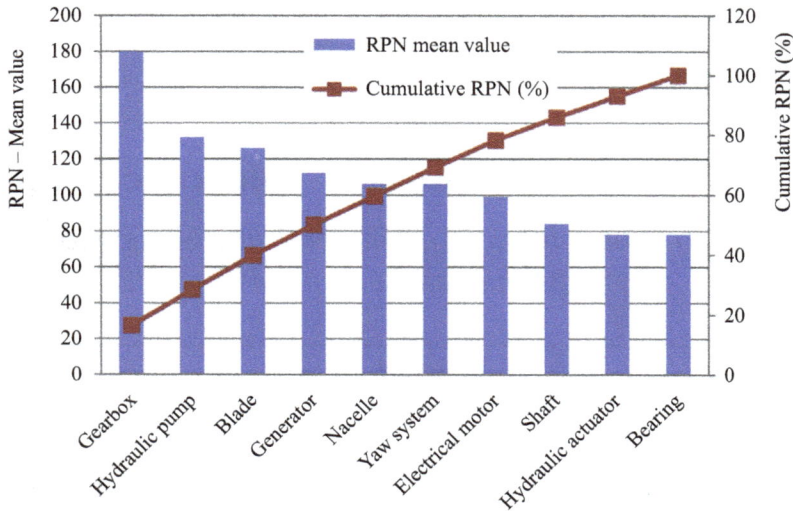

Figure 3.1   Tidal turbine components criticality (RPN stands for risk priority number, which is the multiplication of occurrence, severity, and difficulty of detection values)

and rotor speed and position sensors could be required. These mechanical sensors are often subject to failure, which requires the implementation of advanced fault–tolerant approaches to enhance the overall systems performance and ensure its availability and equipment safety.

In the sensitive context of an immersed system as a tidal turbine, a resilient control and sensors fault–tolerant MPPT is a key feature aiming to increase the availability and energy conversion efficiency. Tidal turbine control issue has been increasingly explored in the literature, where classical PI controllers have been shown to be very sensitive to faulty conditions [11,12]. This comparative study clearly highlights the need for robust fault-resilient control. A critical state-of-the-art review specifically shows that three control techniques seem having the lead, namely model predictive control [13], active disturbance rejection control [14], and high-order sliding modes [15]. Moreover, sensors fault-resilient approaches have been widely investigated and compared where turbine parameters-based approaches have been proved to be efficient for optimal operation [11].

3.2   Fundamentals on faults resilience

Since marine current turbines (MCT) are recent technologies with very low rate of deployment, there is almost no surveys on potential failure of these kinds of technology. However, since the energy conversion components are the same as wind turbine, it can be reasonably considered that the two systems are affected by equivalent faults and may have similar behavior under abnormal operating conditions. Moreover, since MCT are submerged systems, the degree of criticality of these faults should be higher.

3.2.1 Basic concepts on fault tolerance

Fault is defined as an abnormal condition or defect that affects a system and its sub-systems that cause system malfunctioning and may lead to catastrophic failure. Faults can be categorized as either random fault, which occurs as a result of wear and deterioration or systematic fault, which is often due to the error in the specification of the system and therefore affect all system types. Whereas, a failure is considered as a system state or condition in which the system fails in meeting its desirable or intended operation under specified conditions. A specific device failure mainly occurs when it is used past the limits (mechanical and thermal stresses, electrical loading, etc.) of the design specifications. Faults affecting sensors and actuators induce disturbances in the system and controller loss of effectiveness. These issues can be handled by implementing fault–tolerant control approaches. Sensors faults can be bias, drift, scaling, noise, and hard fault (stuck value from the sensor). Fault in the actuators are lock failure, float failure, runaway or handover, and loss of effectiveness.

Fault tolerance is the ability of a system to fulfil its indented function regardless of faults occurrence. Indeed, system performance and stability are generally affected by the fault occurrence but degradation can be acceptable to a certain extent as long as reliability is guaranteed. Fault–tolerant control approaches can be implemented in production critical systems, which is the case for marine current turbines, in order to ensure the continuity of service, minimize downtime, and decrease the operating and maintenance costs.

3.2.2 Marine current turbines failures

Potential failures of electrical drives are mainly related either to the generator (electrical and mechanical parts), gearbox, power electronics converters, transformers, insulators, cables, etc., as described in Figure 3.2.

- Failure of the power supply and power converters, which comprises passive and active components failure, sensors defects, damaged connectors, and faults

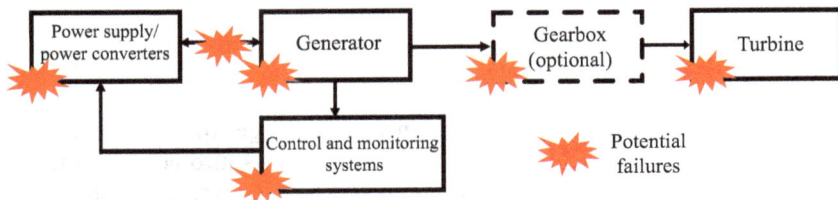

Figure 3.2 Potential failures of electric drives

on the IGBT control signals (short-circuit, opening of the IGBT and abnormal modulations).

- Failure of the electrical generator (bar breakage, insulation breakage, short-circuit between stator conductors, demagnetisation, etc.).
- Failure of the mechanical components (wear or breakage of all or part of the bearings, shafts misalignment, gearbox failure, turbine biofouling or breakage, etc.).
- Failure of the control and monitoring devices.
- Failure of cables, connectors, protection and filtering devices.

These major failures are due to a variety of causes that are associated with design, manufacturing, or improper usage. Failures can have both internal and external causes. Among the origins of electrical drives failures:

- Failures because of internal causes:
  - Mechanical causes include fraction and abrasion, conductors displacement, bearing failures, eccentricity, vibrations which may affect bearings, gearbox, blades, transmission, and support.
  - Electrical causes that concern stator faults (short-circuits, open phase, etc.), insulation faults and rotor faults (demagnetization).
- Failures because of external causes:
  - Mechanical causes include pulsating torque, improper installation.
  - Electrical causes include transient, grid voltage fluctuation, voltage unbalance, harmonics and all power grid disturbances, bad operating regimes (overvoltage, overcurrent, and overspeed).
  - Humans-related causes such as improper sizing, improper assembly, overloading, and lack of maintenance (lubrication, cooling, etc.).
  - Environmental causes include temperature, humidity, biofouling, debris, dust, etc.

These failures have various consequences, which include:

- Ageing, which is due to electrical, mechanical, and thermal stresses.
- Endangering the safety of surrounding equipments and humans.
- Wear, torque oscillations, vibrations, and mechanical fatigue.
- Overheating, risk of short circuit and fire.
- Effects on power grid and end users equipments.
- Acoustic noise.

3.2.3   *Diagnosis and fault tolerance property*

Power systems operating reliability is a constant challenge. Indeed, operating reliability is the set of abilities of a power system that allows it to perform a required function under given conditions. It consists in considering a system not only through its main operating modes but also through the secondary behaviors it may have and for which it was not designed.

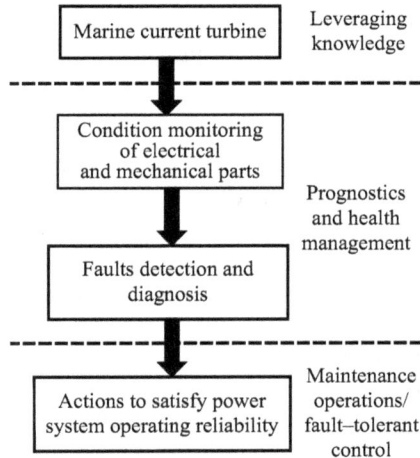

Figure 3.3   Power systems operating reliability design

Operational safety for electrical drives in the context of marine current turbines application is of utmost importance considering that failure of this kind of system is often very expensive in terms of damages caused to the installation, difficulty to repair or replace any components that fail in normal use, and lose of production. The objective is to develop an operating reliability strategy, which allows increased security, reliability, availability, and reduced maintenance and operating costs. This can be achieved at the design stage or by using predictive maintenance and fault–tolerant control as depicted in Figure 3.3:

- Ensuring resilience at the design stage through redundancy (material and analytical redundancy).
- Implementing predictive maintenance, which is mainly based on condition monitoring, faults diagnosis, and remaining useful life estimation.
- Implementing fault–tolerant control (FTC) strategies, which allow accommo- dating faults and maintaining stability with lower or acceptable performance degradation.

3.3   Fault–tolerant systems at design stage

A resilient or fault–tolerant system is a system that has the built-in capability to pre- serve the continued correct execution of its scheduled operations and input/output functions in the presence of a certain set of operational faults. For submerged sys- tem like marine currents turbines, this property should be considered during design stage in order to increase the reliability and availability and decrease the operation and maintenance costs. This can be achieved through hardware and analytical redundancy, multi-phase generators, and reconfigurable power converters using multilevel modular converters (MMC), for instance.

3.3.1   Redundancy

Redundancy is a fundamental aspect in the design of fault–tolerant systems. It is defined as a duplication of critical subsystems such as sensors, actuators, embedded control systems, and functions of the considered safety-critical system. It is intended for increasing the system reliability, enhancing performance, and ensuring the continuity of service. Two functions of redundancy can be distinguished, which are passive redundancy and active redundancy. Passive redundancy uses extra capacity to ensure the margin of safety and lower the impact of failing components. However, when a limited number of faults occur, performance degradation is experienced. One common form of passive redundancy in marine current turbine context is polyphase synchronous generators or double-star asynchronous generators. Unlike passive redundancy, active redundancy uses conditional monitoring of system components, whose output is used in a voting logic. The voting logic allows an automatic components reconfiguration in order to eliminate system performance decline. In MCT application, this is mainly implemented for sensors, control, and communication networks.

3.3.1.1   Hardware redundancy

Redundancy has been originally developed to deal with safety-critical systems hardware components faults. Three or more components (often sensors) are operated in parallel. A voter mechanism is used to consolidate the information and consequently determining the most likely reading as depicted in Figure 3.4. In this figure, the redundancy is termed as static redundancy and requires $2n + 1$ modules to tolerate n faults. For example, triple modular redundancy (TMR) is one of the most reliable hardware redundancies that allow tolerating one module fault. Consequently, the faulty module must be replaced immediately after the occurrence of a fault in the corresponding channel. Otherwise, if two channels are simultaneously faulty, the system shuts down and does not produce output.

Unlike static hardware redundancy, in dynamic redundancy, two parallel modules are sufficient to tolerate a fault. This reduces the number of parallel modules required to ensure hardware redundancy. However, a fault management mechanism must be integrated to perform a fault detection and isolation algorithm as given in Figure 3.5. Indeed, after the fault detection, remedial strategies are used to isolate the faulty components and replace them through automatic changeovers. Two standby strategies can be implemented, which are cold standby as depicted in Figure 3.9(a) and hot standby as given in Figure 3.9(b).

3.3.1.2   Analytical redundancy

Analytical redundancy is based on an observer that provides an estimate of the signals of interest instead of having multiple sensors that measure the same quantity. Indeed, upon the loss of a sensor, it is possible to reconstruct its measurement by a process mathematical model, which is fed by other still active sensors as shown in Figure 3.6.

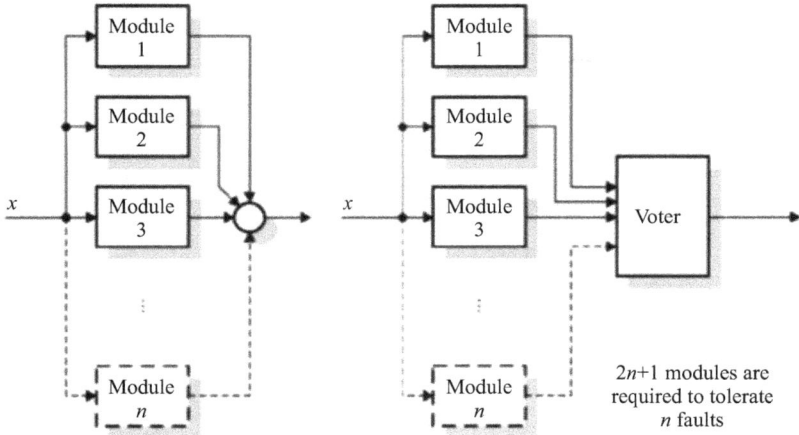

Figure 3.4 Static hardware redundancy

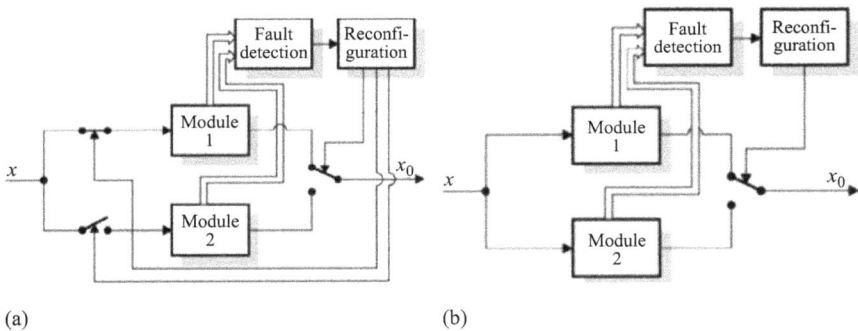

(a)　　　　　　　　　　　　　　　　(b)

Figure 3.5 Dynamic hardware redundancy. (a) Cold standby. (b) Hot standby.

In analytical redundancy configuration, physical sensor measurement is compared with the estimated value computed by the observer and residual is generated. Consequently, abnormal increase in this residual indicates a fault in the sensor. In some circumstances, the output of the observer can be used to control the system until a corrective maintenance is performed to replace the defected sensor. The observer is an algorithm that is implemented in the control embedded system and no additional hardware devices are required. This method can be mainly used to compensate sensors output in case of failure in the field of marine current turbines.

3.3.2 Intrinsic resilient systems

Some components of MCT can be designed to withstand severe conditions and to accommodate some faults. Multiphase generators can be used such as double-star asynchronous generator or multiphase permanent magnets synchronous generator,

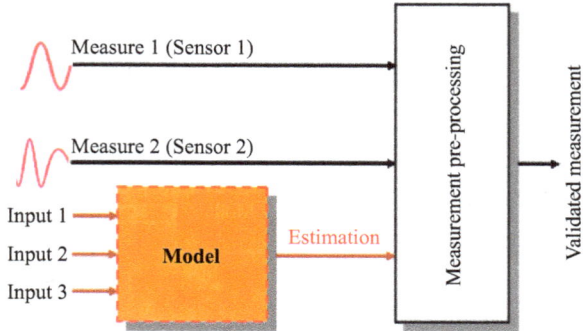

Figure 3.6   Analytical redundancy

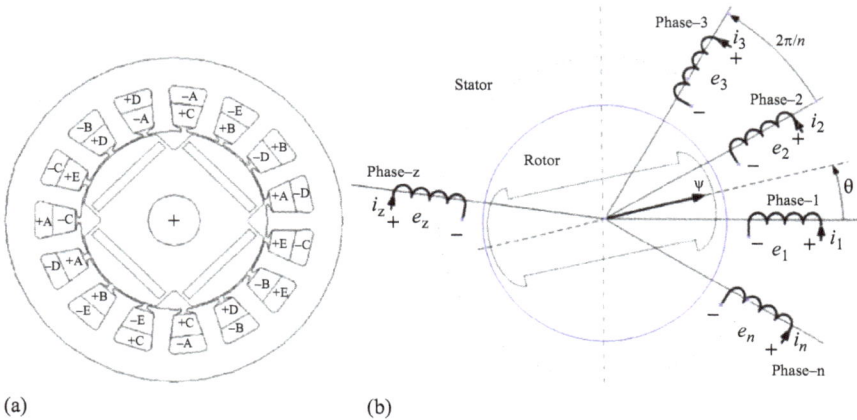

(a)   (b)

Figure 3.7   Multiphase permanent magnet generator. (a) Five phase PMSG.
(b) Cross-sectional view of the geometry of multiphase generator.

which allow the continuity of operation in case of faulty phase. Furthermore, power processing modules can have a redundant legs allowing to tolerate some faults on the power electronics switches.

3.3.2.1   Multi-phase generators

Multiphase generators offer additional degrees of freedom that can be used for resilient operation [16]. Indeed, under faulty conditions, the remaining healthy phases can be used to compensate the faults and continue the tidal turbine operation. Figure 3.7 shows five-phase permanent magnet synchronous generator that has been proposed in the literature for handling a single electrical fault occurrence [15].

3.3.2.2 Power electronics physical reconfiguration

A fault resilient power converter topology is presented in Figure 3.8. In this topology, a physical reconfiguration is possible by using fuses to isolate the faulty leg and triac to connect the corresponding phase to DC link mid-point. Indeed, fault in power electronic switches in a specific inverter leg is first detected and isolated by means of

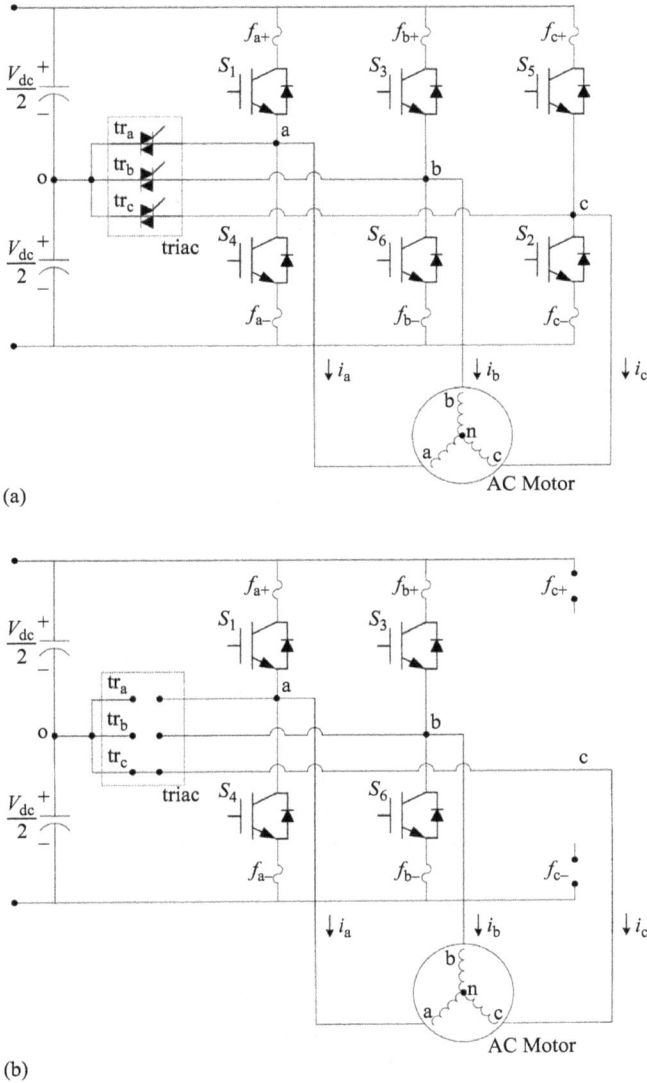

(a)

(b)

Figure 3.8   Reconfigurable power electronics for MCT: topology 1. (a) Pre-fault configuration. (b) Post-fault configuration.

fast-acting fuse and the motor is then connected to mid-point of DC link by turbing-on the triac of associated faulty inverter leg.

Figure 3.9 shows power converter used to fed an AC machine with physical reconfiguration capability for fault tolerance. In this second resilient power converter topology, a fourth redundant leg is used, which is connected to the neutral point through a triac. In case of fault, the faulty inverter leg is first isolated using fuses. Then, the machine neutral point is connected to the fourth redundant leg by turning-on the triac tr_n to accommodate the fault.

3.4 Fault–tolerant control

Fault–tolerant control systems are classified into two main categories: active FTC and passive FTC [17]. An active FTC system requires a fault detection and isolation module and controller reconfiguration. In the opposite, passive FTC system is based on a robust controller to certain disturbances and parameters uncertainties. A hybrid FTC can be implemented allowing to take advantage of the pros of both active and passive FTCs approaches.

3.4.1 Active fault–tolerant control system

The structure of active fault–tolerant controller is depicted in Figure 3.10. It consists of fault detection and isolation (FDI) module, which is generally an observer-based mechanism. An FDI unit allows detecting any deviation of the actual system output from the observer output, which is considered as a faulty condition. Faults can affect both actuators, sensors, and the controlled plant. FDI detects the fault, isolate the failing components and provides online information to the controller to react. A reconfiguration mechanism allows reconfiguring the controller to deal with the new operating conditions. It is worth to notice that several controllers should be implemented at once that allow to cope with specific fault in specific component while maintaining stability an avoiding shut-down or catastrophic failure. The operator is warned about the system malfunction and the faulty components in order carry out the required corrective maintenance operations.

In marine current turbine context, Figure 3.11 gives an example of active TFC system implementation for generator-side converter control. Several controllers are implemented to deal with the possible faults in the system such as direct and indirect vector controllers, sensorless vector or scalar controllers and open loop V/Hz controller. The control decision strategy allows choosing the appropriate controller depending on the information provided by the fault detection and diagnosis module. This module provides information on the operating conditions of system and detects any fault on the turbine, generator (mechanical and electrical parts), or sensors. Another aspect of active FTC is the controller change of operation as depicted in Figure 3.12. An internal model of the system computes an additional controller output that is based on measurements on the actual system. This controller output allows accommodating some faults.

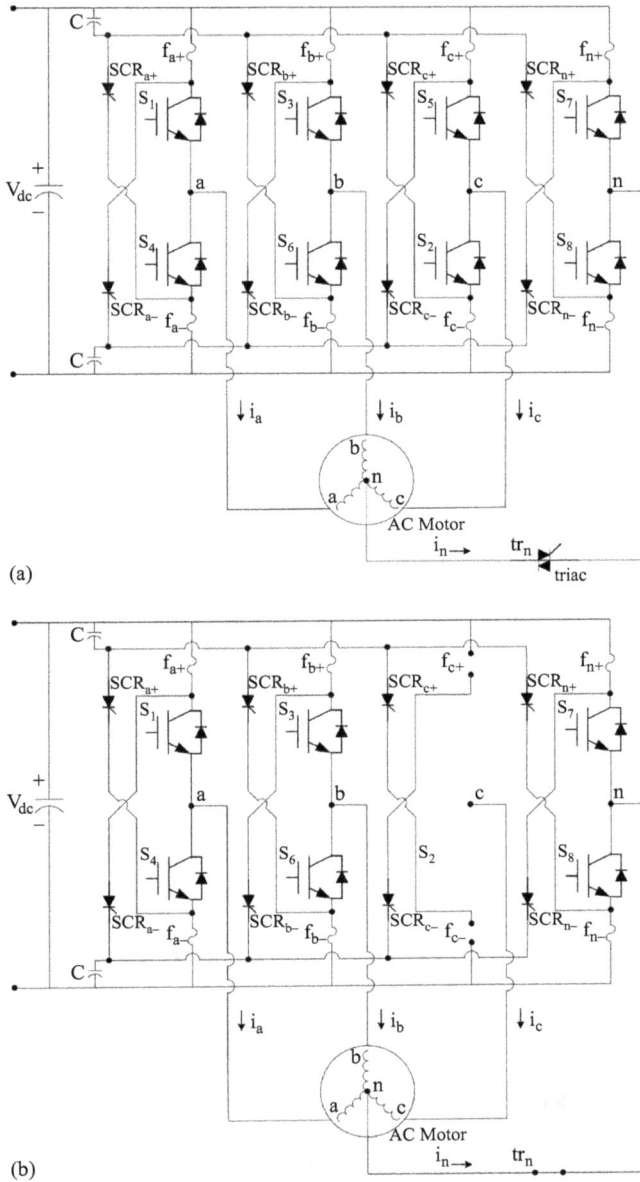

(a)

(b)

Figure 3.9   Reconfigurable power electronics for MCT: topology 2. (a) Pre-fault configuration. (b) Post-fault configuration.

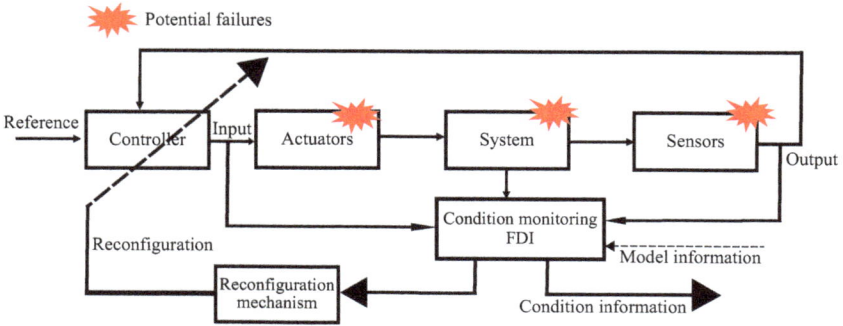

Figure 3.10   Architecture of active fault tolerant control

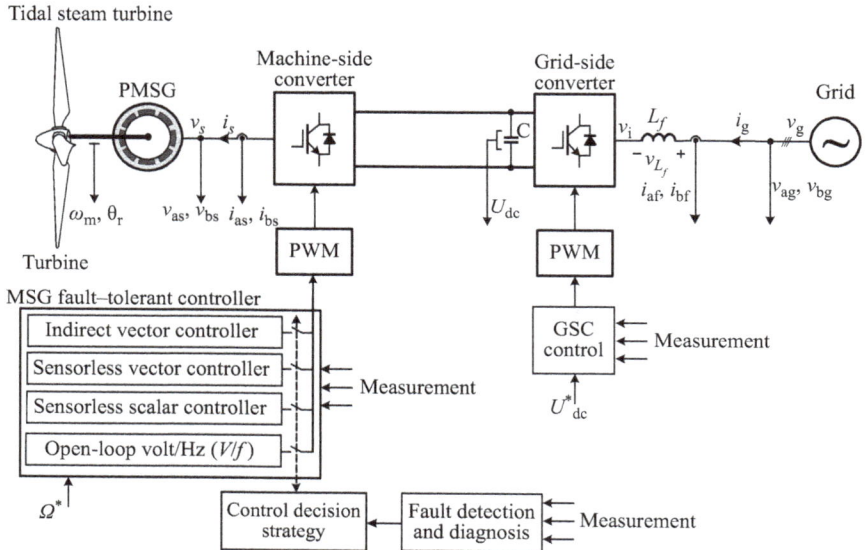

Figure 3.11   Resilient control strategies example – reconfiguration

Active fault–tolerant control presents the advantage of dealing with several faults type as they have been considered at the controller design stage. Moreover, it results in optimal performance and manages to achieve stability and continuity of production. Unfortunately, this approach suffers from several disadvantages and limitations are as follows such as it cannot deal with a large number of fault scenarios and unforeseen faults. Moreover, AFCTS are more complex to implement and more difficult to design for nonlinear systems with uncertainties, their computational cost is high (slow time response), and real-time implementation is a huge challenge for real-time decision-making. Finally, AFCTS depends on the FDI module results, which is sensitive to noise and may lead to false alarm.

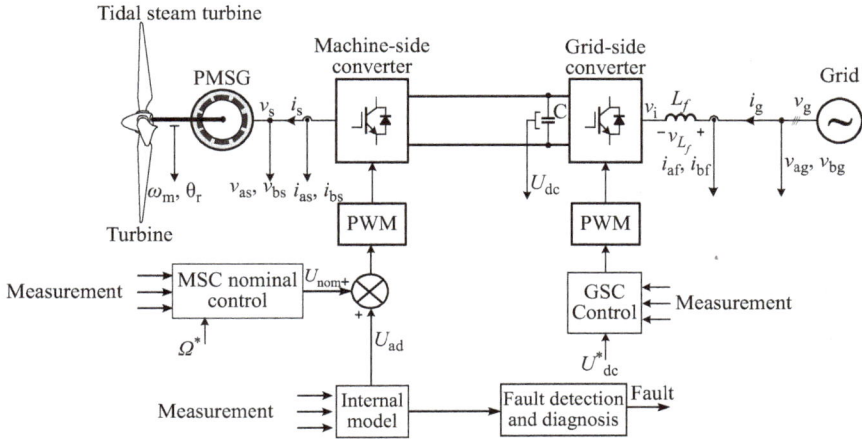

Figure 3.12   Resilient control strategies example – change of operation

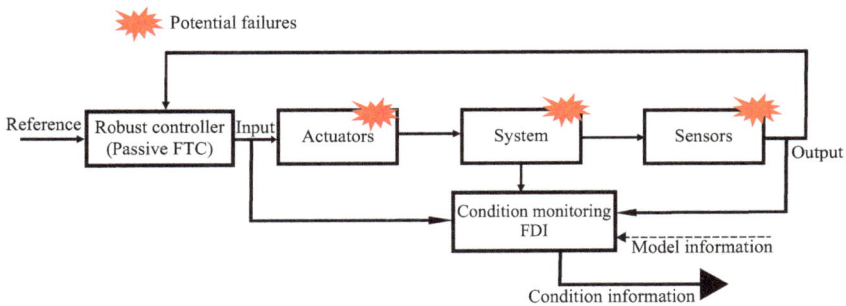

Figure 3.13   Structure of passive fault–tolerant control

3.4.2   Passive fault–tolerant control system

As compared with AFCTS, the architecture of passive fault–tolerant control system (PFTCS) is quite simple as it is mainly based on robust controller to internal/external disturbances and system parameters uncertainties. Figure 3.13 gives the structure of passive fault–tolerant controller. In PFTCS, there is no FDI and no controller recon-figuration is required. The controller predefined parameters are unchanged and allow masking the fault. Sliding mode controller is one of the most popular PFTC since it is robust to external disturbances, system parameters variations and model uncertainties.

Passive fault–tolerant control presents the following advantages:

- Simple to implement and fast time response.
- Robust fixed structure controller.
- Faults have been considered at the controller design stage.
- To certain extent, it can deal with unforeseen faults.

Table 3.1   Comparison of FTC approaches [17]

System property	AFCTS	PFTCS
Architecture	Complex	Simple
Time response	Slow	Fast
Fault detection	Online/real time	Offline
Computational burden	High	Relatively small
Fault detection and isolation	Essential	Required
Controller reconfiguration	Required	Not required
Noise effect	Wrong decision may be made due to noise corruption	Robust to noise
Time delay	Yes due to reconfiguration mechanism	No time delay
Faults nature	Wide variety of faults	Predefined faults
Control structure	Variable	unchanged

Disadvantages and limitations are as follows:

• Controller hide the fault, which can evolve to catastrophic failure.
• System integrity can be jeopardized for complex and several kinds of simultaneous faults.
• Difficult to account for a large number of fault scenarios.

3.4.3   Hybrid fault–tolerant control system

A brief comparison of AFCTS and PFTCS is provided in Table 3.1 based on the FTC architecture complexity, computational burden and immunity to noise and various disturbances. Based on these pros and cons, hybrid FTC can be designed to benefit from the advantages of both previously discussed approaches and allow handling fast disturbances with optimal performance. Specifically, a hybrid FTC system can use multiple controllers, which can handle any incorrect decision from FDI unit, ensure system operation stability, and deal with parameters uncertainties. The system should undergo low and acceptable performance degradation.

3.5   Tidal stream turbines magnets fault–tolerant control case study

This section presents a case study of magnets fault–tolerant control strategy. The demagnetization process, illustrated in Figure 3.14 [18], is mainly due to temperature variations, high stator windings opposing magnetomotive forces (faulty current, Figure 3.14(a)), high torque during transients, and magnets aging [19–21]. Moreover, when partial demagnetization occurs, the same load torque will be generated at the expense of a higher current than in the healthy case. This will consequently increase the temperature constraint accelerating therefore the demagnetization process (Figure 3.14(b)).

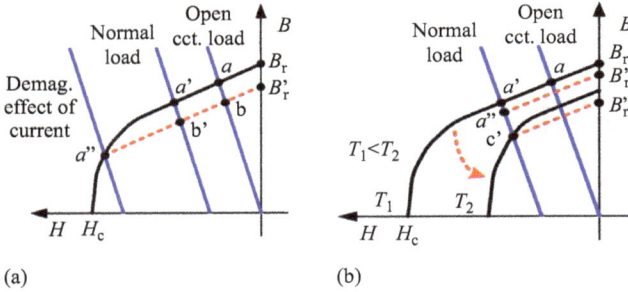

Figure 3.14   Permanent magnets demagnetization. (a) Demagnetizing magnetomotive force. (b) Temperature effect.

The main objective of this research topic is therefore proposing a fault-resilient control approach for a tidal turbine system experiencing a non-critical magnets failure. For that purpose, the magnetic equivalent circuit method will be first used for magnet failures modeling. Conventional PI controllers resiliency limitations are then illustrated. Finally, a second-order sliding mode control strategy is proposed and demonstrated to be robust and efficient for tidal turbine resilience under magnet failures.

3.5.1   Tidal turbine overview

Figure 3.15 illustrates the basic parts of a tidal turbine system. It mainly consists of turbine rotor, permanent magnet synchronous generator coupled to a DC-bus via a three-phase converter. The connection to the grid is achieved based on a three-phase converter (inverter). The generator-side converter (rectifier) is controlled based on the maximum power point tracking (MPPT) method. Since the considered turbine rotor system is non-pitchable, the MPPT is achieved by adjusting the turbine power coefficient at its optimal value C_{pmax} by adjusting the tip speed ratio at its optimal value λ_{opt}. The grid-side converter is controlled to maintain the DC link voltage constant assuming it operates with a unity power factor.

3.5.1.1   Resource

Resource modeling has been previously investigated in [22]. Therefore, knowing the tide coefficient for a specific site, it is easy to derive the following first-order model for tidal current velocities v_t:

$$v_t = v_{nt} + \frac{(C - 45)(v_{st} - v_{nt})}{95 - 45} \tag{3.1}$$

where C corresponds to the tide coefficient, v_{st} and v_{nt} are the spring and neap tide current velocities, respectively.

The tidal current velocity could however be disturbed by long-length oceanic waves, which have a narrow spectrum and much energy (swell effect) [23]. This study

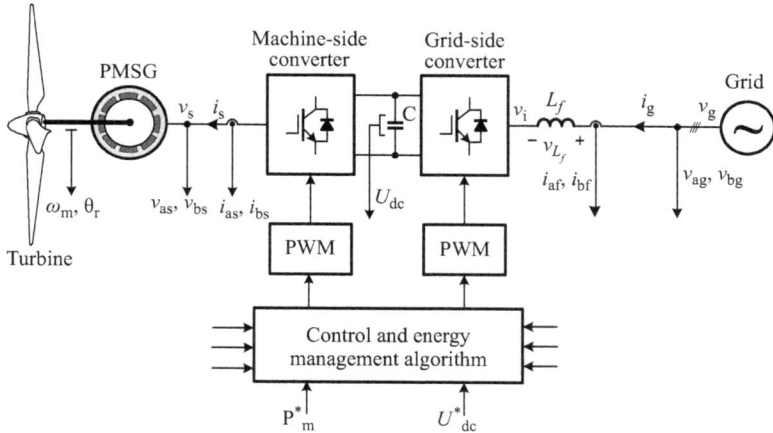

Figure 3.15 Tidal turbine system basic structure

will not consider the swell effect and uses the first-order model in (3.1) to calculate the tidal velocity for each second. Indeed, when using a low speed permanent magnet generator (direct-drive option), the rotor speed variation in a sampling period is very small compared to other variables. Therefore, the rotor speed is assumed unchanged at the next sampling instant.

3.5.1.2 Tidal turbine characteristics

A tidal turbine harnessed power is expressed by

$$P_m = \frac{1}{2}C_p(\lambda, \beta)\rho\pi r^2 v_t^3 \tag{3.2}$$

where $C_p(\lambda, \beta)$ corresponds to the power coefficient. For typical tidal turbines, the maximum value of C_p for normal operation is estimated to be in the range of 0.35–0.5. As for wind turbines, the C_p can be approximated by an equation depending on the tip speed ratio $\lambda = r\Omega/v_t$ and the blade pitch angle β.

A non-pitchable tidal turbine has been adopted, as it is the case of some industrial projects (OpenHydro and Voith Hydro). Indeed, fixed blades enable avoiding an additional relative movement in water, therefore increasing reliability [24]. In this context, the C_p is supposed to be only depending on λ, where Figure 3.16 illustrates the C_p curve used for simulations. For the MPPT strategy implementation purposes, the machine-side converter must be controlled in order to keep the C_p at its maximum value $C_{pmax} = 0.45$, which corresponds to the optimal tip speed ratio $\lambda_{opt} = 6.3$. Several MPPT strategies have been proposed and implemented in the literature such as [25]:

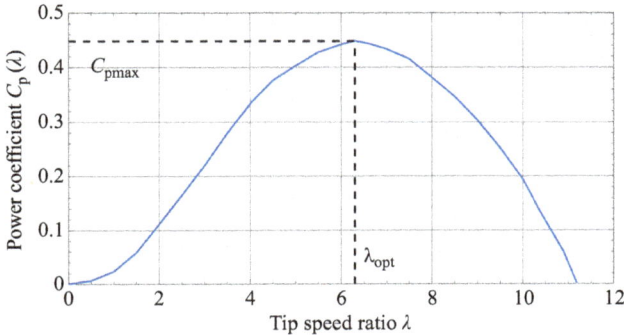

Figure 3.16   Tidal turbine power coefficient curve

- *MPPT by optimal tip speed ratio*: Generator speed is adjusted to keep tip-speed ratio constant at its optimum value λ_{opt} for all marine current speeds. For this MPPT strategy, both current speed and generator speed must be measured.
- *MPPT by turbine profile*: MPPT profile, which is provided by the manufacturer, delivers corresponding active power reference. Measurement of current speed is required.
- *MPPT by optimal torque control*: Torque reference is computed based on the knowledge of the turbine characteristics, particularly maximum power coefficient C_{Pmax} and optimal tip-speed ratio λ_{opt}. To perform torque control, generator speed must be measured.
- *MPPT by perturb and observe*: This strategy does not require current speed or generator speed measurement. The output electrical power is measured and generator speed is adjusted empirically in order to seek for the maximum output power. This approach is generally applied for small-scale wind turbines.

Figure 3.17 illustrates the extractable power for different tidal speeds, while showing two different operating zones. The MPPT zone, where the extractable power is maximized, remains below the generator-rated power. The second zone corresponds to the one where the extractable power is limited to the generator-rated one. Several approaches can be used to limit the extractable power as described in [24]. Figure 3.18 illustrates the permanent magnet synchronous generator-based tidal turbine electromechanical torque for different tidal speed. In the following, the extractable power will be supposed to be always below the rated one. Therefore, only the MPPT zone is investigated.

3.5.2   Magnetic equivalent circuit model for magnets failure modeling

Fault-related signatures rising in PMSM electromotive force (EMF) spectra are mainly due to air-gap flux density changes that are created by magnet defects. Consequently,

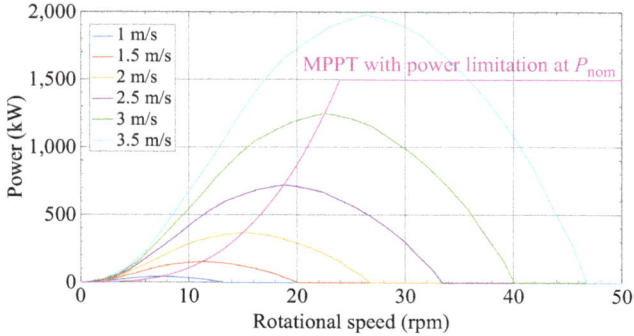

Figure 3.17 Tidal turbine power characteristics versus tidal speed

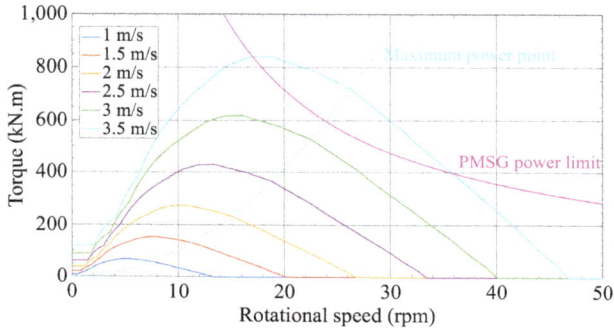

Figure 3.18 Tidal turbine torque characteristics versus tidal speeds

stator flux linkages due to each individual magnet is separately calculated using magnetic equivalent circuit model. Then, the net flux linkage is obtained by an iterative process for superposition: the net stator phase "a" flux linkage is obtained as the cumulative flux linkage of each coil created by each magnet.

3.5.2.1 Stator flux linkage

The flux linkage of a stator coil is the integral of the flux that passes through the air-gap and links to this particular coil, i.e.,

$$\Phi_{mcoil}(\theta_m) = Nl_{st}r_g \int_{\theta=-\frac{\theta_{PM}}{2}}^{\theta=\frac{\theta_{PM}}{2}} B_g(\theta, \theta_m)d\theta \tag{3.3}$$

where N, l_{st}, and r_g correspond to the number of turns of each coil, motor stack length, and air-gap radius, respectively. B_g is the function of the air-gap flux density with respect to the stator space angle θ and electrical rotor position θ_e and is given by

$$
B_g = \begin{cases} B_{g1} = k_{sl}(\theta)\dfrac{\Phi_m}{A_{g1}}, & \theta_e - \dfrac{\theta_{PM}}{2} \leq \theta \leq \theta_e + \dfrac{\theta_{PM}}{2} \\[2ex] B_{g2} = k_{sl}(\theta)\dfrac{\Phi_m}{A_{g2}}, & \text{elsewhere} \end{cases}
\tag{3.4}
$$

where B_{g1} is the flux density over one PM, B_{g2} is the flux density of the rest of the air-gap considering the PMs as air, and $A_{g1} = r_g l_{st}\theta_{PM}$ and $A_{g2} = r_g l_{st}(2\pi - \theta_{PM})$ are the corresponding areas. The flux density Φ_m is given by

$$
\Phi_m = \frac{l_{st}r_g\theta_{PM}\left(1 - \dfrac{\theta_{PM}}{2\pi}\right)}{1 + \mu_{rPM}\dfrac{g}{l_{PM}} + (\mu_{rPM} - 1)\dfrac{\theta_{PM}}{2\pi}} B_r
\tag{3.5}
$$

where B_r corresponds to the PM residual flux density, l_{PM} is the PM length, θ_{PM} is the PM surface angle, μ_{rPM} is the PM relative permeability, and g is the air-gap length. Parameter $k_{sl}(\theta)$ stands for slot correction factor, which models the effect of the stator slots and is given in [26].

Now that the flux linkage of stator coils due to each magnet is separately computed, the magnet fault can be introduced by changing the flux amplitude of the corresponding failing magnet.

3.5.2.2 Magnets failure modeling

The magnetic equivalent circuit method is based on the generator EMF. As the EMF mirrors the magnets flux [27], magnet failure effects will consequently impact the EMF. It could therefore be considered as an interesting mean for the generator magnet failures analysis. The EMF is given by:

$$
e = \frac{d\Phi_{mcoil}}{dt} = \frac{d\theta_e}{dt}\frac{d\Phi_{mcoil}}{d\theta_e} = \omega_e \frac{d\Phi_{mcoil}}{d\theta_e}
\tag{3.6}
$$

For the purpose of magnet demagnetization analysis, each magnet of one coil EMF is calculated using Fourier series [9]:

$$
\begin{cases} e_{a1PM1} = \sum_{n=-\infty}^{+\infty} E_n e^{jn\theta_m} \\[2ex] e_{a1PM2} = \sum_{n=-\infty}^{+\infty} -E_n e^{jn\frac{360}{2N_p}} e^{jn\theta_m} \\[1ex] \vdots \\[1ex] e_{a1PM2N_p} = \sum_{n=-\infty}^{+\infty} -E_n e^{jn(2N_p-1)\frac{360}{2N_p}} e^{jn\theta_m} \end{cases}
\tag{3.7}
$$

where e_{azPMm} is the EMF due to the magnet number m in the zth coil of phase a, and E_n is the EMF amplitude. The zth coil EMF calculated by summing all EMFs due to each magnet, using the superposition rule, is given by

$$e_{az} = \sum_{n=-\infty}^{+\infty} \left(\sum_{i=1}^{2N_p} (-1)^{i-1} k_i e^{jn(i-1)\frac{360}{2N_p}} \right) e^{jn\theta_{cz}} E_n e^{jn\theta_m} \tag{3.8}$$

where the demagnetization effect is introduced by means of the k_i coefficient and $\theta_{cz} = (z-1)\frac{C}{N_s}360$ is the phase a zth coil angle [27]. With N_s is the stator slots number and $C = \frac{N_s}{2N_p}$ is its coil pitch.

After the zth coil EMF calculation, a one-phase EMF is then calculated by the summation of the EMF in all coils. Consequently, phase a EMF is given by:

$$e_a = \sum_{n=-\infty}^{+\infty} \left(\sum_{i=1}^{2N_p} (-1)^{i-1} k_i e^{jn(i-1)\frac{360}{2N_p}} \right) \left(\sum_{z=1}^{N_{coil}} (-1)^{z-1} e^{jn\theta_{cz}} \right) E_n e^{jn\theta_m} \tag{3.9}$$

It can be rewritten as follows:

$$e_a = \sum_{n=-\infty}^{+\infty} k_{mn} k_{san} E_n e^{jn\theta_m} \tag{3.10}$$

Where

- $E_n e^{jn\theta_m}$ represents the EMF due to one magnet in one coil.
- $k_{mn} = \left(\sum_{i=1}^{2N_p} (-1)^{i-1} k_i e^{jn(i-1)\frac{360}{2N_p}} \right)$ is known as the magnet factor. It expresses the magnet failure percentage by the k_i index. Typically, in healthy conditions, k_i is equal to 1 and for all values of the n, k_{mn} will be equal to zero except when the number of pole pairs is odd. It should be mentioned that at failure occurrence, all the mechanical frequency multiples can appear in the EMF waveform [28],
- $k_{san} = \left(\sum_{z=1}^{N_{coil}} (-1)^{z-1} e^{jn\theta_{cz}} \right)$ is termed the coil factor in each phase. It expresses the location and the number of coils. For other phases, the amplitude of this factor is the same. However, the phase-shift is equal to 120°.

The other phases will have an identical EMF but with a different coil factor as follows:

$$\begin{cases} e_b = \sum_{n=-\infty}^{+\infty} k_{mn} k_{sbn} E_n e^{jn\theta_m} \\ e_c = \sum_{n=-\infty}^{+\infty} k_{mn} k_{scn} E_n e^{jn\theta_m} \end{cases} \tag{3.11}$$

According to the aforementioned equations, k_{mn} is the coefficient that corresponds to the fault-related harmonics and is common in all three phases and the coil factor k_{sxn} corresponds to the phase difference between stator phase variables.

3.5.3   Sliding mode-based magnet failure-resilient control

Sliding mode control (SMC) is a non-linear control method that has been widely investigated in the literature due to its inherent robustness and relative easiness of design and implementation for real-time applications. SMC is the control methods that are robust to modeling uncertainties and external disturbances. SMC is based on discontinuous control approach, which allows to rapidly switch from one continuous manifold to another. The system dynamics are forced to predefined location in the state space called the sliding surface. First-order sliding mode (FOSM) control suffers from chattering phenomena due to imperfections of switching devices that lead to oscillatory dynamic behavior near the sliding surface. To overcome this issue, a high order (HOSM) has been proposed that allows chattering reduction in a finite-time [29,30]. In fact, computer simulation results show that the higher order sliding mode controller succeeds in reducing the chattering phenomenon that is present in the first-order controller. Chattering may result in excessive heating of power electronics and wear of mechanical components.

Consider the following non-linear system described in the state space as follows:

$$\dot{x}(t) = f(x(t)) + g(x(t))\, u(t) \tag{3.12}$$

where $x(t)$ is the state variable vector, $f(.)$ and $g(.)$ are smooth vector fields in the state space, and $u(t)$ is the discontinuous control given by

$$u(t) = \begin{cases} U^+ & s(x,t) > 0 \\ U^- & s(x,t) < 0 \end{cases} \tag{3.13}$$

where $s(.)$ corresponds to the chosen sliding surface to ensure that the state variables track their desired trajectories. The control inputs U^+ and U^- can be either scalar values or state $x(t)$ dependent. The system described by (3.12) shows sliding mode properties when:

- The reachability is met: reachability condition ensures that the trajectory of the system state variables will approach and eventually reach the sliding surface. Mathematically speaking, this condition is formulated as

$$s(x)\dot{s}(x) < 0 \tag{3.14}$$

- The existence condition should be met: existence condition ensures that once the trajectory is within the vicinity of the sliding surface, it is still always directed toward the sliding surface, which implies

$$\lim_{s \to 0} s\dot{s} < 0. \tag{3.15}$$

- The stability condition is met: it ensures that the sliding surface always directs the trajectory toward a stable equilibrium (steady-state stability).

The state-space model of the surface-mounted permanent magnets synchronous machine ($L_{sd} = L_{sq} = L_s$), supplied with balanced three-phase voltage sources ($v_{s0} = 0$), is as follows:

$$\begin{cases} \dfrac{di_{sd}}{dt} = -\dfrac{R_s}{L_s}i_{sd} + N_p\Omega i_{sq} + \dfrac{v_{sd}}{L_s} \\[2mm] \dfrac{di_{sq}}{dt} = -\dfrac{R_s}{L_s}i_{sq} - N_p\Omega\left(i_{sd} + \dfrac{\Phi_m}{L_s}\right) + \dfrac{v_{sq}}{L_s} \\[2mm] \dfrac{d\Omega}{dt} = \dfrac{\Gamma_{em}}{J_t} - \dfrac{f_v\,\Omega + \Gamma_s}{J_t} \end{cases} \tag{3.16}$$

Unmodeled dynamics of PMSM can lead to parameter uncertainties. In the following, this model's parameters' variation is giving as follows:

$$R_s = \hat{R}_s + \Delta R_s \quad L_s = \hat{L}_s + \Delta L_s \quad J_t = \hat{J}_t + \Delta J_t$$

$$f_v = \hat{f}_v + \Delta f_v \quad \Phi_m = \hat{\Phi}_m + \Delta\Phi_m \quad \Gamma_s = \hat{\Gamma}_s + \Delta\Gamma_s$$

where ΔX is a bounded disturbance of the parameter X. The sliding surfaces for the three state variables are defined as:

$$\begin{cases} S_d = i_{sd} - i_{sd}^* = 0 \\ S_q = i_{sq} - i_{sq}^* = 0 \\ S_\Omega = \Omega - \Omega^* = 0 \end{cases} \tag{3.17}$$

3.5.3.1   First-order sliding mode control design

Direct and quadrature axis control design
The direct and quadrature voltages computed based on first-order sliding mode control are composed of two components, namely the equivalent control $v_{sx,eq}$ and the on–off control $v_{sx,N}$ as follows:

$$\begin{cases} v_{sd} = v_{sd,eq} + v_{sd,N} \\ v_{sq} = v_{sq,eq} + v_{sq,N} \end{cases} \tag{3.18}$$

The equivalent control is computed by solving $\dot{S}_d = 0$ and $\dot{S}_q = 0$. The derivative of the chosen sliding surface is given by

$$\begin{cases} \dot{S}_d = \dfrac{i_{sd}}{dt} - \dfrac{i_{sd}^*}{dt} = \dfrac{1}{L_s}\left[-R_s i_{sd} + N_p\Omega L_s i_{sq} + v_{sd} - L_s\dfrac{i_{sd}^*}{dt}\right] \\[3mm] \dot{S}_q = \dfrac{i_{sq}}{dt} - \dfrac{i_{sq}^*}{dt} = \dfrac{1}{L_s}\left[-R_s i_{sq} - N_p\Omega(L_s i_{sd} + \Phi_m) + v_{sq} - L_s\dfrac{i_{sq}^*}{dt}\right] \end{cases} \tag{3.19}$$

Solving $\dot{S}_x = 0$ leads to the following expression of the equivalent control for d and q-axis:

$$\begin{cases} v_{sd,eq} = \hat{R}_s i_{sd} - N_p \Omega \hat{L}_s i_{sq} + \hat{L}_s \dfrac{i^*_{sd}}{dt} \\[3mm] v_{sq,eq} = \hat{R}_s i_{sq} + N_p \Omega \left(\hat{L}_s i_{sd} + \Phi_m \right) + \hat{L}_s \dfrac{i^*_{sq}}{dt} \end{cases} \tag{3.20}$$

Therefore, $\dot{S}_d$ and $\dot{S}_q$ can be rewritten as

$$\begin{cases} \dot{S}_d = \dfrac{1}{L_s} \left[-\Delta R_s i_{sd} - \Delta L_s \left(\dfrac{i^*_{sd}}{dt} - N_p \Omega i_{sq} \right) \right] + \dfrac{1}{L_s} v_{sd,N} \\[4mm] \dot{S}_q = \dfrac{1}{L_{sq}} \left[-\Delta R_s i_{sq} - N_p \Omega (\Delta L_s i_{sd} + \Delta \Phi_m) - \Delta L_s \dfrac{i^*_{sq}}{dt} \right] + \dfrac{1}{L_s} v_{sq,N} \end{cases} \tag{3.21}$$

Due to boundedness of the uncertainties present in the PMSG model parameters, there exists a positive constant v_{d0} and v_{q0} such that

$$\begin{cases} v_{d0} > \left| \Delta R_s i_{sd} + \Delta L_s \left(\dfrac{i^*_{sd}}{dt} - N_p \Omega i_{sq} \right) \right| \\[4mm] v_{q0} > \left| \Delta R_s i_{sq} + N_p \Omega (\Delta L_s i_{sd} + \Delta \Phi_m) + \Delta L_s \dfrac{i^*_{sq}}{dt} \right| \end{cases} \tag{3.22}$$

Consequently, the switching component of the control signal is defined as:

$$\begin{cases} v_{sd,N} = -v_{d0}\, sgn(S_d) \\ v_{sq,N} = -v_{q0}\, sgn(S_q) \end{cases} \tag{3.23}$$

where the sign function $sgn(.)$ is defined as

$$sgn(S) = \begin{cases} 1 & S(x) > 0 \\ 0 & S(x) = 0 \\ -1 & S(x) < 0 \end{cases} \tag{3.24}$$

Finally, the control action that aims at keeping the direct and quadrature currents at their desired reference values is given by:

$$\begin{cases} v_{sd} = \left[\hat{R}_s i_{sd} - N_p \Omega \hat{L}_s i_{sq} + \hat{L}_s \dfrac{i^*_{sd}}{dt} \right] - v_{d0}\, sgn(S_d) \\[4mm] v_{sq} = \left[\hat{R}_s i_{sq} + N_p \Omega \left(\hat{L}_s i_{sd} + \Phi_m \right) + \hat{L}_s \dfrac{i^*_{sq}}{dt} \right] - v_{q0}\, sgn(S_q) \end{cases} \tag{3.25}$$

Direct and quadrature current control design based on rotational speed dynamics

The d-axis current reference is maintained equal to zero, i.e. $i^*_{sd} = 0$, in order to minimize the copper losses, therefore minimizing currents for a given torque. In this case, the PMSG torque can be directly controlled through the q-axis current. The quadratic current reference i_{sq}^* is calculated by the speed controller. Let us define the quadrature axis current reference control law as follows:

$$i_{sq}^* = i_{sq,eq}^* + i_{sq,N}^* \tag{3.26}$$

where $i_{sq,eq}^*$ corresponds to the equivalent control and $i_{sq,N}^*$ denotes the switching control. Based on the sliding surface for the rotational speed defined in (3.17), the sliding surface derivative is found to be

$$\dot{S}_\Omega = \frac{3}{2} \frac{N_p \Phi_m}{J_t} i_{sq} - \frac{f v \Omega}{J_t} - \frac{d\Omega^*}{dt} \tag{3.27}$$

Hence, the equivalent control is given by

$$i_{sq,eq}^* = \frac{1}{K_t} \left[\hat{f}_v \Omega + \hat{J}_t \frac{d\Omega^*}{dt} \right] \tag{3.28}$$

where $K_t = \frac{3}{2} N_p \Phi_m i_{sq}$.

Following the same methodology as previously discussed for on–off component control design, there exists a positive constant i_{q0} such that:

$$i_{q0} > \left| \frac{1}{K_t} \Delta \hat{f}_v \Omega + \Delta \hat{J}_t \frac{d\Omega^*}{dt} \right| \tag{3.29}$$

Consequently, the switching component of the control signal is defined as:

$$i_{sq,N}^* = -i_{q0} \, sgn(S_\Omega) \tag{3.30}$$

Finally, the quadrature current reference is given by

$$i_{sq}^* = \frac{1}{K_t} \left[\hat{f}_v \Omega + \hat{J}_t \frac{d\Omega^*}{dt} \right] - i_{q0} \, sgn(S_\Omega) \tag{3.31}$$

3.5.3.2   High-order sliding mode control design

To reduce the chattering effect, high order sliding mode control methods have been proposed in the literature. The derivation of higher order sliding mode (HOSM) control using the super-twisting algorithm is presented hereafter [31]. The sliding surfaces for HOSM control are chosen the same as the FOSM control.

Direct and quadrature axis control design

For second-order sliding mode control, both $S(x)$ and $\dot{S}(x)$ should be zero. Hence, one more derivative of each function must be computed. Based on the results of the previous section, the second derivative of

$$\begin{cases} \ddot{S}_d = -\frac{R_s}{L_s} \frac{d i_{sd}}{dt} + N_p \frac{d\Omega}{dt} i_{sq} + N_p \Omega \frac{d i_{sq}}{dt} + \frac{1}{L_s} \frac{d v_{sd}}{dt} - \frac{d^2 i_{sd}^*}{dt^2} \\ \\ \ddot{S}_q = -\frac{R_s}{L_s} \frac{d i_{sq}}{dt} - N_p \frac{d\Omega}{dt} \left(i_{sd} + \frac{\Phi_m}{L_s} \right) - N_p \Omega \frac{d i_{sd}}{dt} + \frac{1}{L_s} \frac{d v_{sq}}{dt} - \frac{d^2 i_{sq}^*}{dt^2} \end{cases} \tag{3.32}$$

Let us denote

$$
\begin{cases}
\psi_d = -\dfrac{R_s}{L_s}\dfrac{di_{sd}}{dt} + N_p\dfrac{d\Omega}{dt}i_{sq} + N_p\Omega\dfrac{di_{sq}}{dt} - \dfrac{d^2 i_{sd}^*}{dt^2} \\[4mm]
\psi_q = -\dfrac{R_s}{L_s}\dfrac{di_{sq}}{dt} - N_p\dfrac{d\Omega}{dt}\left(i_{sd} + \dfrac{\Phi_m}{L_s}\right) - N_p\Omega\dfrac{di_{sd}}{dt} - \dfrac{d^2 i_{sq}^*}{dt^2}
\end{cases}
\tag{3.33}
$$

and

$$
\begin{cases}
\eta_d = \dfrac{1}{L_s}\dfrac{dv_{sd}}{dt} \\[4mm]
\eta_q = \dfrac{1}{L_s}\dfrac{dv_{sq}}{dt}
\end{cases}
\tag{3.34}
$$

with $\psi_d \in [-\Psi_d,\ \Psi_d]$, $\psi_q \in [-\Psi_q,\ \Psi_q]$, $\eta_d \in [\mathcal{N}_m,\ \mathcal{N}_M]$ and $\eta_q \in [\mathcal{N}_m,\ \mathcal{N}_M]$.

The control input consists of the sum of two components: the equivalent control $v_{sd,eq}$ and $v_{sq,eq}$ found in the previous section and $\tilde{v}_{sd}$ and $\tilde{v}_{sq}$, which are defined as

$$
\begin{cases}
\tilde{v}_{sd} = v_{sd1} + v_{sd2} \\[2mm]
\tilde{v}_{sq} = v_{sq1} + v_{sq2}
\end{cases}
\tag{3.35}
$$

where:

$$
\begin{cases}
\dot{v}_{sd1} = -W_d sgn(S_d) \\[2mm]
\dot{v}_{sq1} = -W_q sgn(S_q)
\end{cases}
\tag{3.36}
$$

with,

$$
\begin{cases}
W_d > \dfrac{\Psi_d}{\mathcal{N}_m} \\[4mm]
W_q > \dfrac{\Psi_q}{\mathcal{N}_m}
\end{cases}
\tag{3.37}
$$

and

$$
v_{sd2} = \begin{cases}
-\lambda_d\,|S_0|^p\, sgn(S_d) & |S_d| > |S_0| \\[2mm]
-\lambda_d\,|S_d|^p\, sgn(S_d) & |S_d| \leq |S_0|
\end{cases}
\tag{3.38}
$$

$$
v_{sq2} = \begin{cases}
-\lambda_q\,|S_0|^p\, sgn(S_q) & |S_q| > |S_0| \\[2mm]
-\lambda_q\,|S_q|^p\, sgn(S_q) & |S_q| \leq |S_0|
\end{cases}
\tag{3.39}
$$

where

$$
\lambda_d^2 \geq \frac{4\Psi_d}{\mathcal{N}_m^2}\frac{\mathcal{N}_M}{\mathcal{N}_m}\frac{W_d + \Psi_d}{W_d - \Psi_d} \quad
\lambda_q^2 \geq \frac{4\Psi_q}{\mathcal{N}_m^2}\frac{\mathcal{N}_M}{\mathcal{N}_m}\frac{W_q + \Psi_q}{W_q - \Psi_q}
\tag{3.40}
$$

and

$$
0 < p \leq 0.5
\tag{3.41}
$$

The choice $p = 0.5$ leads to the maximum real sliding order for 2-sliding realization (real-sliding order 2 is achieved). Thus,

$$\begin{cases} \tilde{v}_{sd} = -\lambda_d \, |S_d|^{\frac{1}{2}} \, sgn(S_d) - W_d \int sgn(S_d)dt \\ \tilde{v}_{sq} = -\lambda_q \, |S_q|^{\frac{1}{2}} \, sgn(S_q) - W_q \int sgn(S_q)dt \end{cases} \tag{3.42}$$

The overall second-order sliding mode control laws for the direct and quadrature currents are as follows:

$$\begin{cases} v_{sd} = v_{sd,eq} - \lambda_d \, |S_d|^{\frac{1}{2}} \, sgn(S_d) - W_d \int sgn(S_d)dt \\ v_{sq} = v_{sq,eq} - \lambda_q \, |S_q|^{\frac{1}{2}} \, sgn(S_q) - W_q \int sgn(S_q)dt \end{cases} \tag{3.43}$$

Quadrature current control design based on rotational speed dynamics
From the computation of the previous section, the second derivative of the rotational speed sliding surface is as follows:

$$\ddot{S}_\Omega = \frac{K_t}{J_t} \frac{di_{sq}}{dt} - \frac{f_v}{J_t} \frac{d\Omega}{dt} - \frac{d^2\Omega^*}{dt^2} \tag{3.44}$$

Let us denote

$$\begin{cases} \psi_\Omega = -\frac{f_v}{J_t} \frac{d\Omega}{dt} - \frac{d^2\Omega^*}{dt^2} \\ \eta_\Omega = \frac{K_t}{J_t} \frac{di_{sq}}{dt} \end{cases} \tag{3.45}$$

with, $\psi_\Omega \in [-\Psi_\Omega, \, \Psi_\Omega]$ and $\eta_\Omega \in [\mathcal{N}_m, \, \mathcal{N}_M]$.

The overall quadrature current reference is the sum of the equivalent control input determined in the previous section and $\tilde{i}_{sq}^*$ is given by

$$i_{sq}^* = i_{sq,eq}^* + \tilde{i}_{sq}^* \tag{3.46}$$

where,

$$\tilde{i}_{sq}^* = i_{sq1} + i_{sq2} \tag{3.47}$$

These two components are given by

$$\dot{i}_{sq1} = -W_\Omega sgn(S_\Omega) \quad i_{sq2} = \begin{cases} -\lambda |S_0|^p \, sgn(S_\Omega) & |S_\Omega| > |S_0| \\ -\lambda |S_\Omega|^p \, sgn(S_\Omega) & |S_\Omega| \leq |S_0| \end{cases} \tag{3.48}$$

To ensure the finite-time convergence to the sliding manifold, the sufficient conditions are:

$$\begin{cases} W_\Omega > \dfrac{\Psi_\Omega}{\mathcal{N}_m} \\ \lambda_\Omega^2 \geq \dfrac{4\Psi_\Omega}{\mathcal{N}_m^2} \dfrac{\mathcal{N}_M}{\mathcal{N}_m} \dfrac{W_\Omega + \Psi_\Omega}{W_\Omega - \Psi_\Omega} \\ 0 < p \leq 0.5 \end{cases} \tag{3.49}$$

The overall second-order sliding mode control law for the quadrature current reference i_{sq}^* is as follows:

$$i_{sq}^* = i_{sq,eq}^* - \lambda_\Omega |S_\Omega|^{\frac{1}{2}} \, sgn(S_\Omega) - W_\Omega \int sgn(S_\Omega)dt \qquad (3.50)$$

It should be mentioned that sliding modes main drawback is chattering due to the usage of the "sign" switching function. The chattering issue is greatly attenuated by the use of higher sliding modes based on the super-twisting algorithm. Indeed, this is due to two simultaneous actions:

- the integral action that filters high frequencies of the chattering phenomena;
- the multiplying terms λ_d, λ_q, W_d, and W_q that become small close to zero.

This is clearly shown in [30] and experimentally validated in [29,32]. The stability proof has not been explicitly considered in this application-oriented study. Indeed, it has been already considered in details by Fridman *et al.* [31]. In this relevant work, Fridman has clearly proved that the stability is guaranteed for second-order sliding modes.

3.5.4   Simulation results

The simulated PMSG-based tidal turbine parameters are given in Table 3.2. For fixed-pitch turbines, the maximum output energy can be achieved if the turbine rotor operates at the optimal regimes characteristic, which requires the tip speed ratio to be optimal. Hence, the rotational speed must be tracking its reference, which depends on the tidal speed for achieving maximum power point tracking. Figure 3.19 represents an example of tidal velocity for a day (each second represents an hour) for a specific location at the Raz de Sein in Bretagne, France (Figure 3.20). It can be seen that a 2.3 m/sec tidal speed peak can be reached.

Two control strategy families are evaluated for PMSG-based marine current turbines: classical PI controller and second-order sliding mode control. The sensitivity of PI controllers to magnets failure is illustrated both on rotor speed, PMSG torque, and output power. Then, the obtained results with second-order sliding mode control are presented and seem to be encouraging regarding magnets-failure resilience capability.

3.5.4.1   PI controllers limitations

The classical PI controller is widely used in industry applications, its design procedure is simple enough, requiring little feedback signals and providing an easy to be implemented control strategy. Tidal turbine simulation under magnets failure has been first carried out using PI controllers, where the failure has been introduced at $t = 2$ sec.

In this context, Figure 3.21 depicts the generator rotor speed under different values of the coefficient k (5% and 10%) in order to evaluate the impact of magnets failure severity on the tidal turbine behavior. It can be clearly noticed speed ripples as soon as the magnet failure appears. These ripples obviously increase with the

Table 3.2   Permanent magnet synchronous generator-based
tidal turbine parameters

Subsystem	Parameter	Value
Tidal turbine	Radius	8 m
	Number of blades	3
	Fluid density	$1,027.68$ kg/m^3
PMSG	Rated power	1.5 MW
	Rated speed	25 rpm
	Stator resistance	0.0081 Ω
	d-axis inductance	1.2 mH
	q-axis inductance	1.2 mH
	Permanent magnets flux	2.458 Wb
	System total inertia	1.3131×10^6 kg m^2
	Viscosity coefficient	8.510^{-3} Nm/s
Converter	Turn-on time	0.13 μs
	Turn-off time	0.445 μs
	Dead time	4 μs
	Duty-cycle frequency	5 kHz

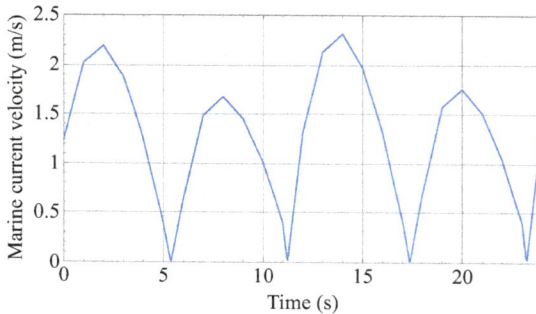

Figure 3.19   Tidal current velocity

demagnetization (factor k increases). At this stage, it seems that a classical PI control is unfortunately not able to efficiently tolerate magnet failures.

PI controller is well-known for its good performance for constant references and for guaranteeing a constant tracking error when the reference is a ramp. Its gains are computed by compensating the larger system time constant. However, in case of a marine current turbine, the generated reference signals are time-varying and tidal velocities are time-varying due to turbulence and swell effect. This makes PI controller not efficient. Moreover, in the case of PMSG magnets failure, it is demonstrated that rotor speed experiences huge ripples, which may destroy the energy conversion system. Consequently, there is a need for a more reliable, secure, and fault–tolerant

Figure 3.20   Tidal turbine expected installation site (B)

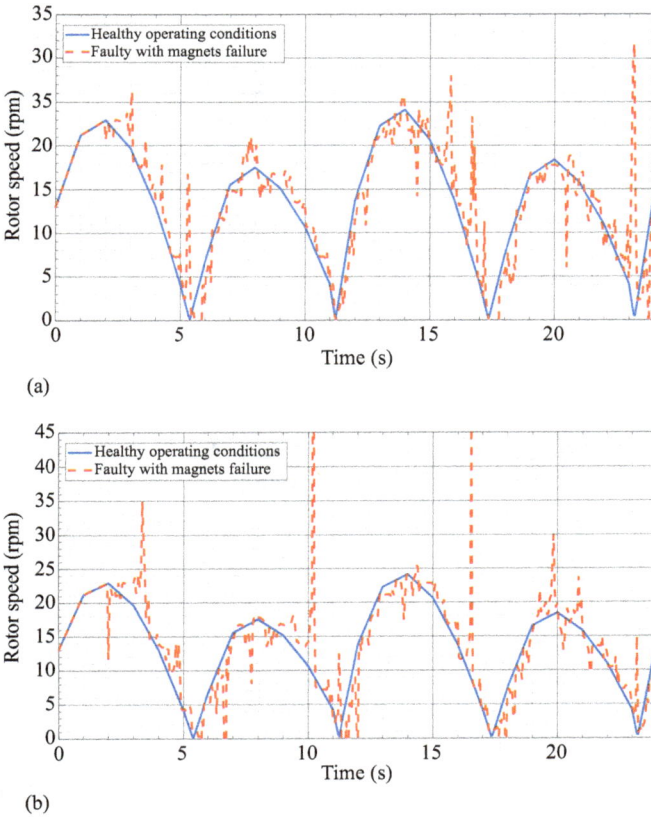

(a)

(b)

Figure 3.21   Tidal turbine simulation under magnets failure. (a) Generator rotor speed (k = 5%). (b) Generator rotor speed (k = 10%).

control strategy, which behaves better than PI controllers. Specifically, there is a clear need for more robust control techniques such as the second-order sliding mode.

3.5.4.2 Proposed approach performance

Sliding mode control (SMC) methods have been adopted for magnets failure resilient control in marine current turbine context. Figure 3.22 shows the overall control scheme. The dynamics of the PMSM and the SMC control methods are implemented in MATLAB®/Simulink®. The generator-side converter is controlled based on SMC methods and grid-side converter is still controlled based on PI controllers. Several studies in the literature have shown that SMC methods effectively reject the parameter variations and external disturbances. While both the first-order and higher order methods offer good performance, these results verify that the higher order sliding mode control method allows eliminating chattering and offers superior performance. Consequently, high-order sliding mode control has been implemented in this study to achieve permanent magnet synchronous generator field-oriented control and an MPPT by optimal tip speed ratio has been implemented for maximizing output power.

The tidal turbine generated power, rotor speed, and torque under faulty conditions are illustrated in Figure 3.23, respectively, with 5% of magnet failure at $t = 2$ sec using PI and second-order sliding mode controls. The achieved results clearly show that demagnetization deteriorates the PI controller performance, while with the second-order sliding mode control, the tidal turbine system power and dynamic performances are almost degradation free. These results obviously confirm that the second-order sliding mode outperforms a PI control in terms of magnet failure resilience effectiveness. Figure 3.24 gives speed, torque, and power ripples variation ranges under magnets failure condition (5% generator demagnetization) for

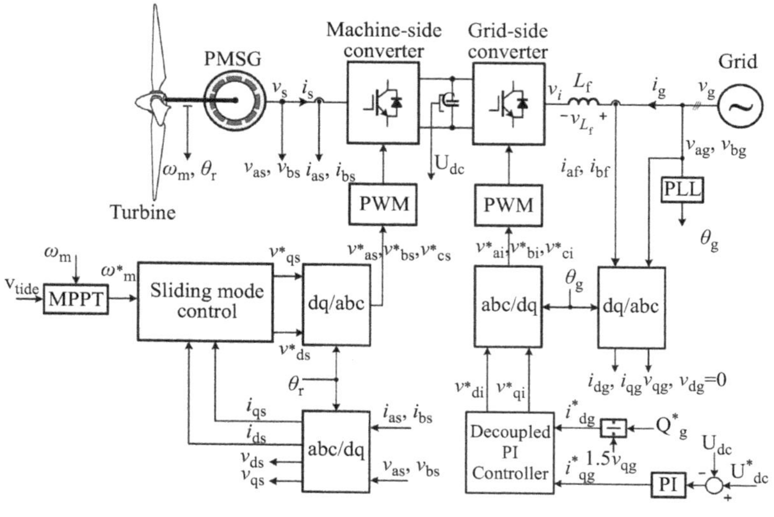

Figure 3.22 Tidal turbine proposed resilient-control structure

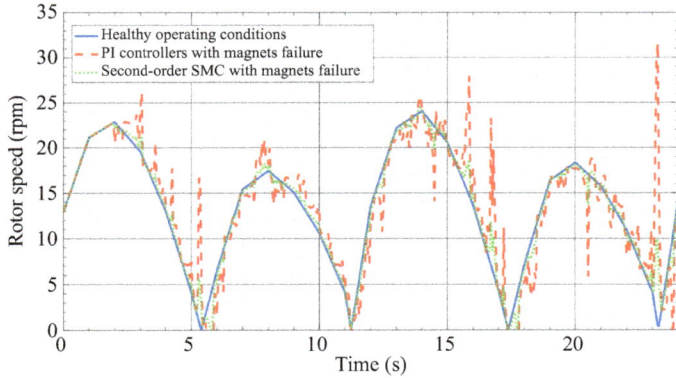

(a)

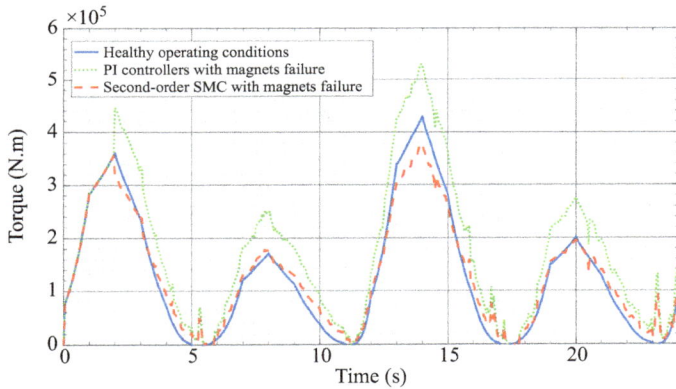

(b)

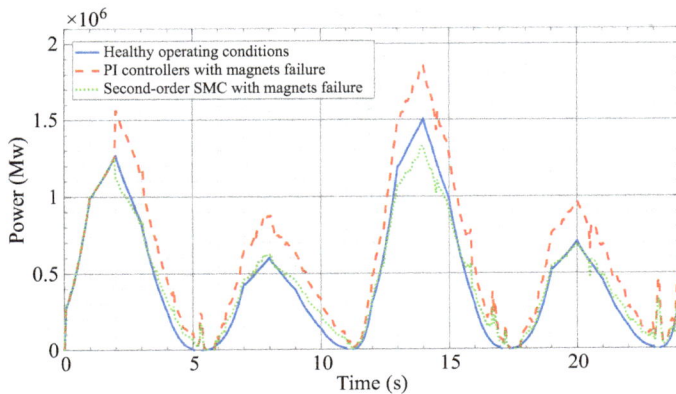

(c)

Figure 3.23 *Tidal turbine generator simulation illustrating fault-resilient control.*
(a) Generator rotor speed. (b) Generator torque. (c) Generated power.

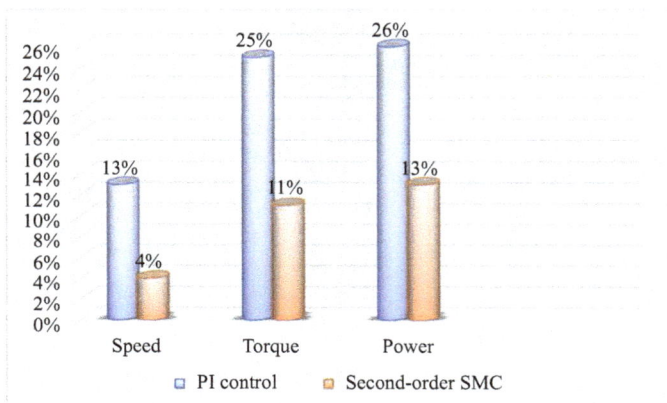

Figure 3.24   Speed, torque, and power variations with 5% demagnetization failure

the two considered controllers. It clearly confirms the appropriateness of the proposed second-order sliding mode control approach for resilience purposes.

These results confirm that the proposed sliding mode control methods are robust to modeling uncertainties and magnets failure. Computer simulation results show that the higher order sliding mode controller succeeds in reducing the chattering phenomenon that is present in the first-order controller. Chattering may result in excessive heating of power electronics and wear of mechanical components. These consequences involve accelerated aging and reduced reliability. Computer simulation studies show the efficacy of the proposed sliding mode control methods applied to a permanent magnet synchronous generator in MCT application.

3.6   Tidal stream turbines sensors fault–tolerant control case study

This section deals with TST sensors failure tolerance. Specifically, a comparative study is presented to assess sensor fault–tolerant control strategies performance in terms of energy conversion efficiency, torque ripple, turbine parameters variations robustness, and computational cost. Several simulations have been carried out on a direct-drive fixed-pitch MCT equipped with a PMSG. These simulations use a real marine current speed data at the Raz de Sein site in Bretagne, France [11].

3.6.1   Flow-meter FTC

To implement a maximum power point tracking (MPPT), the tip–speed ratio must be kept to its optimal value (λ_{opt}) by adjusting the rotational speed of the turbine as follows:

$$\Omega^* = \frac{\lambda_{opt} v_t}{r} \qquad (3.51)$$

To do so, marine current speed is required, which is in general measured using a flow-meters that can be either mechanical or acoustic current meters. These sensors can be lost or degraded due to harsh operational conditions. To ensure TST continuity of service, remedial solutions are implemented, which can be categorized into: turbine parameters-based FTC and turbine parameters-free FTC methods.

3.6.1.1 TST parameters-based FTC methods

These methods require the knowledge of turbine parameters, mainly the power coefficient curve versus the tip–speed ratio. They include power signal feedback control, optimal torque control, and current speed estimation control methods. Their efficiency is highly dependent on the accuracy of turbine model and its parameters identification. Unfortunately, these parameters are affected by several phenomena such as biofouling and ageing.

1. Power signal feedback control (PSF): turbine maximum power curve with respect to rotational speed is given by

$$P_{max} = \frac{\rho \pi r^5 C_{pmax}}{2\lambda_{opt}^3} \Omega^3 = K_{opt} \Omega^3 \tag{3.52}$$

   Consequently, to achieve maximum power extraction, the speed reference is computed as follows:

$$\Omega^* = \sqrt[3]{\frac{P_m}{K_{opt}}} \tag{3.53}$$

2. Optimal torque control (OT): to achieve optimal operation, generator torque is controlled as follows:

$$T_{opt} = \frac{\rho \pi r^5 C_{pmax}}{2\lambda_{opt}^3} \Omega^2 = K_{opt} \Omega^2 \tag{3.54}$$

   This configuration requires the implementation of rotational speed sensor. Note that the rotational speed can be estimated using observers based on the measurement of the electrical quantities (mainly phase currents).

3. Marine current speed estimation control (MCSE): this approach deals with the estimation of marine current speed based on the measurement of the output power and the rotational speed. In [33], authors have proposed the use of Gaussian radial basis function network to estimate the wind speed. In this comparative study, the same approach is implemented to estimate the marine current speed as follows:

$$\hat{v}_t = b + \sum_{j=1}^{h} v_j exp \left(-\frac{\|x - C_j\|^2}{\sigma_j^2} \right) \tag{3.55}$$

   where $x = [P_m, \Omega, \beta]$ is the input vector, $C_j \in \mathbb{R}$ and $\sigma_j \in \mathbb{R}$ are the center and the width of the jth radial basis function unit in the hidden layer, respectively. h is the number of radial basis function units, b and v_j are the bias term and weight between the hidden and output layers, respectively. All these parameters

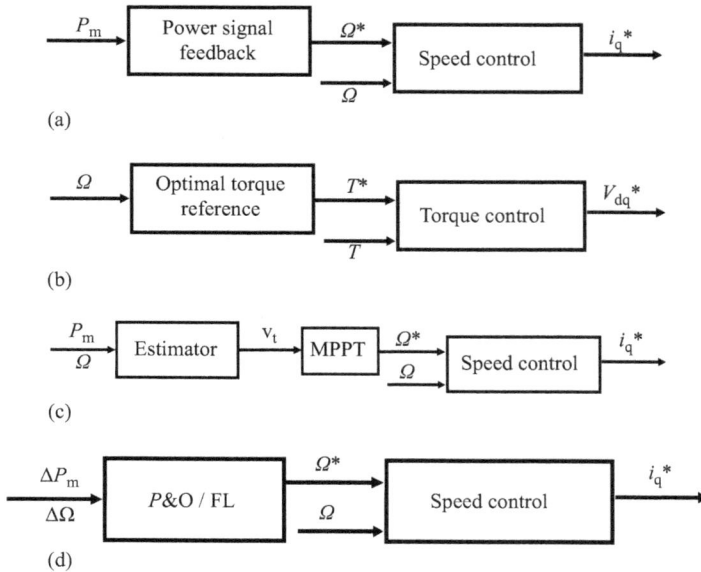

Figure 3.25   Flow-meter FTC methods illustration. (a) PSF control. (b) OT control. (c) MCSE control. (d) (P &O) and Fuzzy logic control.

are determined offline during the training stage using data given by the energy conversion system dynamics.

All these methods scheme is provided in Figure 3.25.

3.6.1.2   TST parameters free-based FTC methods

These approaches' principle is to continuously adjust the turbine rotational speed reference based on the output power in order to achieve a maximum power extraction.

1. Perturb & observe (P&O) control: the search for the optimal operating point is achieved by continuously increasing/decreasing the rotational speed by a constant amount $\Delta\Omega$ based on the measurement of actual output power. This approach is also known as the hill climb search (HCS). The search process continues in the same direction if $\Delta P_e > 0$ and, in the opposite direction, if $\Delta P_e < 0$. The choice of the step size is of paramount importance for this approach. Indeed, a small step size reduces the oscillation around the maximum power point but decreases the convergence speed. Contrariwise, a large step size allows a faster convergence to the maximum power point, but reduces the accuracy of the MPPT algorithm.
2. Fuzzy logic control: it is based on the same principle as P&O control but presents better results as the step size is continuously changing during the maximum power point search. This leads to fast time response and considerably reduces the oscillations around the optimal point. A standard fuzzy logic control strategy requires

three stages: fuzzification, control based on rule table, and defuzzification as described in [34].

3.6.2 *Rotor speed/position sensor FTC*

PMSG rotor position measurement is of paramount importance to implement a field-oriented control or torque control in TST applications. Rotor position can be measured using resolvers or incremental encoders. These sensors can be subjected to failure, which considerably affect the energy conversion efficiency. To solve this issue, a speed estimation can be implemented, which can be achieved by open-loop computation or a closed-loop observers as shown in Figure 3.26.

3.6.2.1 Open-loop computation

This approach allows to estimate the rotor speed or position using electrical variables such as electromotive force (EMF) of magnetic variables such as the flux. Note that

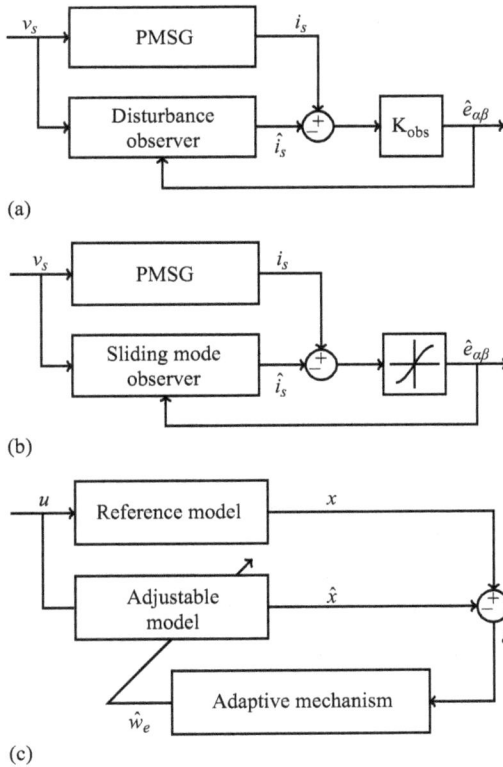

Figure 3.26 *Closed-loop observers methods illustration. (a) Disturbance observer scheme. (b) Sliding mode observer scheme. (c) Model reference adaptive system-based method.*

the later can be estimated from the current and voltage measurements. Flux linkage-based method is one of the most interesting approaches to get the flux, which can be computed as follows [35]:

$$\begin{cases} \phi_{r\alpha} = \int (v_\alpha - R_s i_\alpha)dt - L_s i_\alpha \\ \phi_{r\beta} = \int (v_\beta - R_s i_\beta)dt - L_s i_\beta \end{cases} \tag{3.56}$$

where $\phi_{r\alpha}$ and $\phi_{r\beta}$ are the rotor flux linkages in the $\alpha\beta$ reference frame. Finally, the rotor position can be calculated by

$$\theta_e = \arctan\left(\frac{\phi_{r\beta}}{\phi_{r\alpha}}\right) \tag{3.57}$$

Unfortunately, this approach performance depends on PMSG parameters accuracy and current/voltage transducers measurements reliability. Moreover, this approach is sensitive to noise and internal and external disturbances. Consequently, observers-based approaches have been proposed in the literature to enhance the disturbances rejection capability and improve FTC performance.

3.6.2.2 Disturbance observers

To overcome the aforementioned issues, closed-loop observers are used. Disturbance observers estimate the rotor position and speed based on back EMF, which is computed from PMSG electrical model as follows:

$$\frac{d}{dt}\begin{bmatrix}\hat{i}_\alpha \\ \hat{i}_\beta\end{bmatrix} = \begin{bmatrix} -\dfrac{R_s}{L_s} & 0 \\ 0 & -\dfrac{R_s}{L_s} \end{bmatrix} \times \begin{bmatrix}\hat{i}_\alpha \\ \hat{i}_\beta\end{bmatrix} + \frac{1}{L_s}\left(\begin{bmatrix}v_\alpha \\ v_\beta\end{bmatrix} - \begin{bmatrix}\hat{e}_\alpha \\ \hat{e}_\beta\end{bmatrix}\right) \tag{3.58}$$

and

$$\frac{d}{dt}\begin{bmatrix}\hat{e}_\alpha \\ \hat{e}_\beta\end{bmatrix} = K_{obs}\frac{d}{dt}\begin{bmatrix}\hat{i}_\alpha - i_\alpha \\ \hat{i}_\beta - i_\beta\end{bmatrix} \tag{3.59}$$

where $\hat{x}$ is the estimated value of x and K_{obs} is the observer gain. Finally, the rotor position can be estimated as follows:

$$\hat{\theta}_e = \arctan\left(-\frac{\hat{e}_\alpha}{\hat{e}_\beta}\right) \tag{3.60}$$

3.6.2.3 Sliding mode observers

The principle of sliding mode observer is similar to the disturbance observer. However, the EMF estimation stage is performed using a non-linear function as follows:

$$\begin{bmatrix}\hat{e}_\alpha \\ \hat{e}_\beta\end{bmatrix} = K_{obs}H\begin{bmatrix}\hat{i}_\alpha - i_\alpha \\ \hat{i}_\beta - i_\beta\end{bmatrix} \tag{3.61}$$

where K_{obs} corresponds to the observer gain and $H(.)$ is the sigmoid function defined as

$$H(x) = \frac{2}{1 + e^{-ax}} - 1 \tag{3.62}$$

3.6.2.4 Model reference adaptive system-based methods

This approach is based on two models: the reference model and the adjustable model. The output errors between the two models are used to estimate the rotor speed. This estimate allows to adjust the adjustable model. The reference model is described by the following model in the state space:

$$\frac{dx}{dt} = Ax + u \tag{3.63}$$

where:

- $x = \begin{bmatrix} x_1 \\ x_2 \end{bmatrix} = \begin{bmatrix} i_d + \dfrac{\psi_m}{L_d} \\ i_q \end{bmatrix}$

- $u = \begin{bmatrix} u_1 \\ u_2 \end{bmatrix} = \begin{bmatrix} \dfrac{v_d L_d + R_s \psi_m}{L_d^2} \\ \dfrac{v_q}{L_q} \end{bmatrix}$

- $A = \begin{bmatrix} -\dfrac{R_s}{L_d} & \dfrac{L_q \omega_e}{L_d} \\ \dfrac{L_q \omega_e}{L_d} & -\dfrac{R_s}{L_d} \end{bmatrix}$

The adjustable model is described by

$$\frac{d\hat{x}}{dt} = \hat{A}\hat{x} + u \tag{3.64}$$

where $\hat{x}$ and $\hat{A}$ are the estimated quantities of x and A. Matrix A is continuously adjusted by using the estimated rotor speed as follows:

$$
\begin{aligned}
\hat{\omega}_e &= \int_0^t k_1 \left[i_d \hat{i}_q - i_q \hat{i}_d - \psi_m (i_q - \hat{i}_q) \right] d\tau \\
&\quad + k_2 \left[i_d \hat{i}_q - i_q \hat{i}_d - \psi_m (i_q - \hat{i}_q) \right] + \omega_e(0)
\end{aligned}
\tag{3.65}
$$

3.6.2.5 Extended Kalman filter-based methods

The main objective of the extended Kalman filter is to estimate the state variables $x = \begin{bmatrix} i_\alpha, i_\beta, \Omega, \theta_e \end{bmatrix}^T$. A standard EKF algorithm is composed of two steps:

- Prediction step:
$$x_{k|k-1}^n = x_{k-1|k-1}^n + \left[f(x_{k-1|k-1}^n) + B < u_{k-1}^n > \right] T_c$$

$$P_{k|k-1}^n = P_{k-1|k-1}^n + \left(F_{k-1} P_{k-1|k-1}^n + P_{k-1|k-1}^n F_{k-1}' \right) T_c + Q^n$$
- Innovation step:
$$x_{k|k}^n = x_{k|k-1}^n + K_k^n \left(y_k^n - H x_{k|k-1}^n \right)$$

$P_{k|k}^n = P_{k|k-1}^n - K_k^n H P_{k|k-1}^n$ where the Kalman filter gain matrix is calculated as

$$K_k^n = P_{k|k-1}^n H' \left(H P_{k|k-1}^n H' + R^n \right)'$$

where Q^n and R^n are the variance matrices, P is the estimation error covariance matrix, and T_c is the sampling time.

3.6.3 Comparison criteria

To evaluate the performance of the previously mentioned FTC methods, four criteria are used, which are energy criterion, torque ripples, robustness, and computational cost.

3.6.3.1 Energy criterion

This criterion is used to evaluate the power conversion efficiency of the compared FTC methods. Indeed, all proposed algorithms are supposed to track the maximum power point through the estimation of marine current speed or the rotational speed. It is defined as the ratio between the converted energy during fault conditions and the generated energy for a healthy operating conditions. This can be mathematically expressed as follows:

$$\%E = \frac{\int_{t1}^{t2} P_{FTC} dt}{\int_{t1}^{t2} P_{MPPT} dt} \tag{3.66}$$

3.6.3.2 Torque ripples

Torque ripple can induce TST shaft fatigue and reduce its useful life by accelerating ageing process. Therefore, the main objective while choosing an appropriate FTC method is to minimize torque ripples both in steady-state conditions and during transients. Two factors are considered to evaluate torque ripples:

* Permanent stress during fault:

$$S_p = \sqrt{\frac{\int_{t1}^{t2} (\Delta T - \bar{\Delta T})^2 dt}{t_2 - t_1}} \tag{3.67}$$

where $\Delta T = T_m - T_e$ and t_1 and t_2 are the time instant when sensors are lost and recovered, respectively.
* Stress during switching control law:
  - $S_{t1} = \Delta T_{\max} - \Delta T_{\min}$ computed at t_1, the moment the sensor is disconnected and the FTC strategy applied.
  - $S_{t2} = \Delta T_{\max} - \Delta T_{\min}$ computed at t_2, the moment the sensor is recovered and FTC strategy is disabled.

3.6.3.3 Robustness

FTC methods for both flow-meters and position/speed sensors faults are mainly based on the knowledge of the system model and parameters identification accuracy. In real-world applications, these parameters actual values and the ones used for TFC implementation may differ due to many factors such as temperature and ageing. Therefore, robustness analysis against parameters uncertainties is crucial for optimal FTC design. Hereafter, the parameters variation robustness of the FTC is conducted by assessing the energy loss when there is parameters mismatch between the FTC and the actual system.

3.6.3.4 Computational cost

This criterion evaluates the computational burden for each step of the FTC algorithm to assess its suitability for the implementation in real-world applications. The control law is computed using MATLAB/Simulink with Intel processor i5-2410 with CPU at 2.3 GHz.

3.6.4 *Main simulation results*

The studied FTC strategies have been simulated on 1.5-MW TST as shown in Table 3.2. Two faults are considered, which are flow-meter fault and rotor speed/position transducers failure. Performances are evaluated based on the criteria previously presented.

3.6.4.1 Flow-meter fault case study

Flow-meter failure has been investigated first. At $t_1 = 40$ s, the flow-meter is disconnected and FTC strategies are applied and performance is analyzed and compared. Then, at $t_2 = 80$ s, the sensor is recovered and the system operates in healthy operating conditions. Table 3.3 provides simulation results evaluating the performance analysis criteria. Analyzing this table shows that OT and PSF approaches have less torque ripples and consequently may lead to less TST shaft fatigue. Specifically, PSF method does not lead to torque ripple when switching the control law, which is quite interesting. Figure 3.27 depicts the TST extracted power and energy production. It appears that the FTC methods allow continuity of service and improve availability. Moreover, the methods based on the knowledge of turbine parameters are very efficient. However, for the other methods, the power conversion efficiency is decreasing. Indeed, OT, PSF, and CSE-based approaches allow 99.9% energy conversion efficiency while HCS and FL efficiency is approximately around 80%.

FTC strategies robustness against TST parameters variation is crucial for optimal operation. Figure 3.28 depicts the evolution of the extracted power with respect to errors on the optimal tip–speed ratio λ_{opt}. These results clearly show that unlike parameters-free approaches, FTC parameters-dependent methods performance is significantly affected by errors on λ_{opt} estimation. Indeed, an overestimation or underestimation of λ_{opt} leads to error in the power coefficient that consequently implies a decrease in the energy conversion efficiency as depicted in Figure 3.29. This can be explained using Figure 3.30. In fact, an error on the λ_{opt} estimation leads to error on the turbine optimal rotational speed that causes a decrease on the output electrical power. Consequently, the parameters-dependent FTC methods performance is questionable as its performance hugely depends on the perfect knowledge of the power coefficient curve that may vary during operation due to ageing or biofouling.

Table 3.3   Flow-meter FTC methods performance

Methods	%E	S_p	S_{t1}	S_{t2}	Computational cost
PSF	99.9908%	2.9647e5	508, 730	352, 710	0.26 ms
OT	99.9868%	2.9507e5	No	No	0.26 ms
MCSE	99.9724%	3.7450e5	1, 016, 210	1, 128, 000	0.28 ms
HCS (P&O)	82.8278%	3.4470e5	966, 610	973, 700	0.20 ms
FL	89.1724%	1.3426e5	96, 610	763, 100	0.22 ms

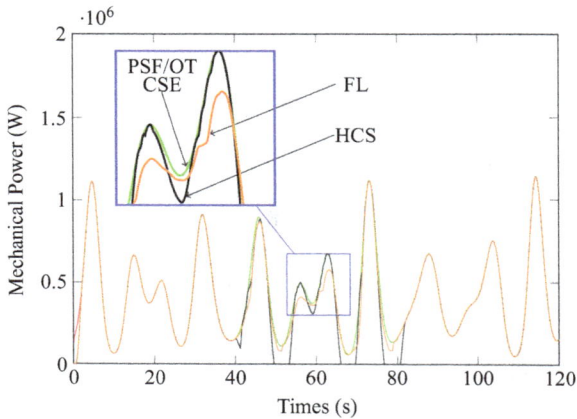

(a)

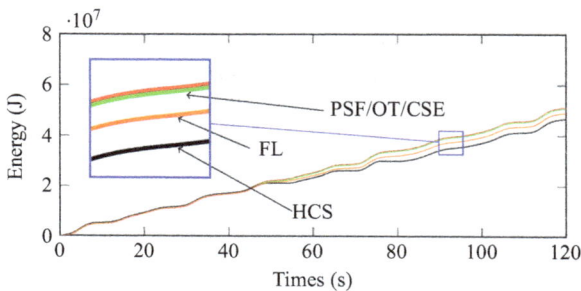

(b)

Figure 3.27   Power and energy extracted from an MCT with flow-meter fault. (a) Mechanical power. (b) Energy generation.

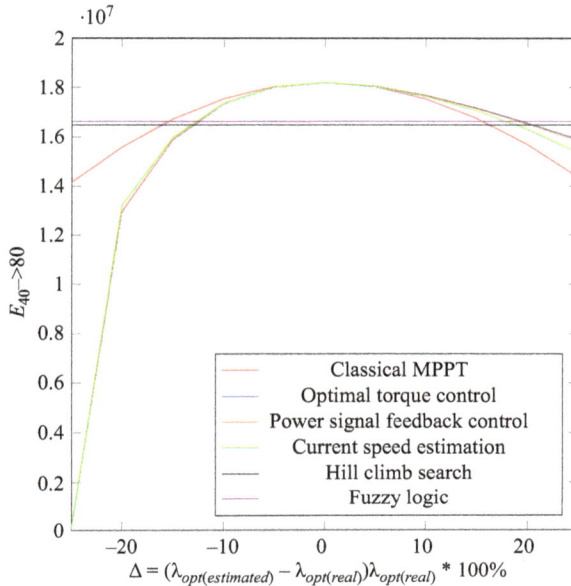

Figure 3.28   Robustness of the studied FTC methods versus errors in the estimated optimal tip–speed ratio λ_{opt}

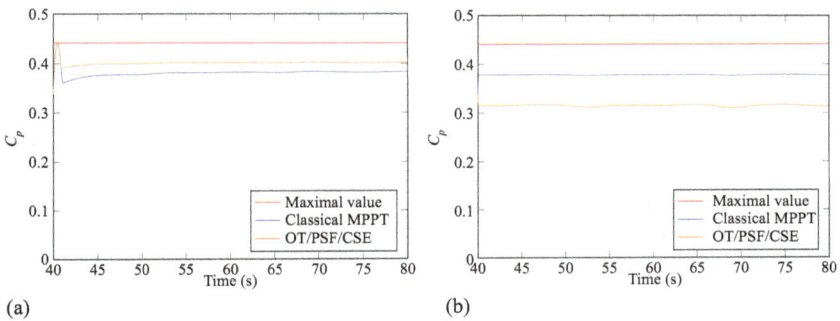

Figure 3.29   Effective power coefficient used for MPPT during flow-meter fault with $\pm 20\%$ errors on the estimation of λ_{opt}

3.6.4.2   Rotor speed/position sensor faults case study

As for the flow-meter fault, speed sensor failure is considered at $t_1 = 40$ s and then the sensor is recovered at $t_2 = 80$ s. Table 3.4 summarizes main simulation results based on the comparison criteria. These results prove the effectiveness of the discussed FTC methods against rotor speed failure as the energy conversion efficiency is greater than

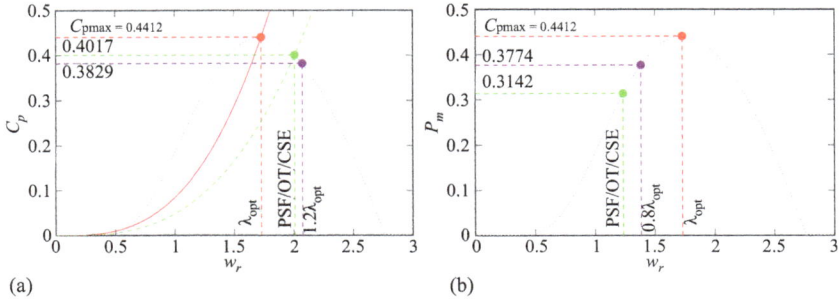

Figure 3.30   Graphical analysis of the actual power extraction under λ_{opt} estimation errors. (a) +20% overestimation. (b) −20% underestimation.

Table 3.4   Rotor speed sensor FTC methods performance

Methods	%E	S_p	S_{t1}	S_{t2}	Computational cost
FLBM	99.9400%	4.2874e5	755, 100	552, 800	0.30 ms
DO	99.9669%	6.0894e5	*no*	127, 470	0.32 ms
SMO	99.9731%	5.4517e5	854, 493	280, 417	0.34 ms
MRAS (P&O)	99.9762%	6.4430e5	No	128, 011	0.32 ms
EKF	99.9732%	3.9210e5	1, 160, 200	116, 310	0.89 ms

99.9%. However, this is performed at the expense of higher torque ripples, which may be dangerous for the TST and surrounding equipment. Figure 3.31 depicts power and energy losses after rotor speed sensor failure. It appears from these curves that flux linkage-based method (FLBM) is the less efficient method as it implies largest power losses.

3.7   Conclusions and perspectives

This chapter presented fault tolerance strategies in the context of marine current turbine. FTC system allows tolerating faults in the MCT while maintaining system stability with low or no performance degradation. In MCT, fault resilience can be implemented at design stage by using polyphase generators, redundant legs power converters, redundant sensors, cables, embedded control systems, and communication networks. Moreover, fault–tolerant control approaches can be implemented with no additional hardware usage. Two approaches have been reviewed and pros and cons discussed: active and passive FTC. Some studies propose the use of hybrid FTC

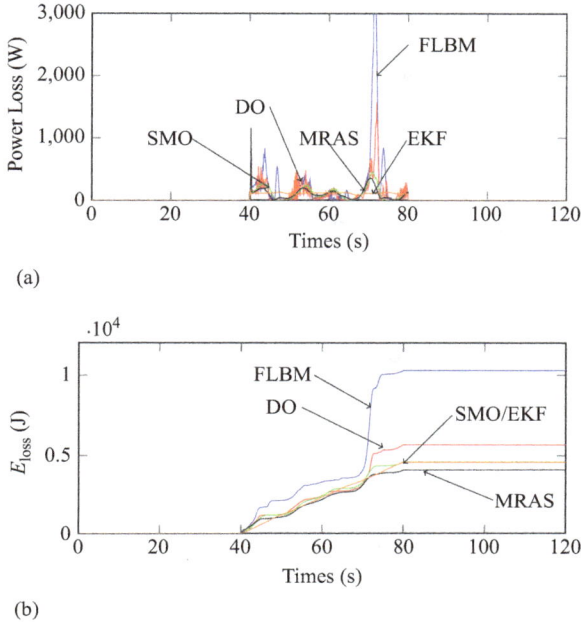

Figure 3.31   Power and energy losses during speed/position sensor failure.

allowing to take advantage the two previously presented methods at the expense of higher complexity and higher computational cost.

A robust fault-resilient control approach for a tidal turbine system experiencing permanent magnets demagnetization case study has been presented. In this context, the magnetic equivalent circuit method has been used for magnet failures modeling in a synchronous generator. The preliminary simulation results have illustrated the tidal turbine power generation and dynamic performances high level of degradation when using conventional PI controllers. Therefore, high-order sliding modes have been used for resilience purposes, while maintaining the tidal turbine optimal power and dynamic performances. Simulations that were carried out with real tidal velocities at the Raz de Sein site in France have clearly shown the second-order sliding mode control advantages and superiority in terms of magnet failure resilience.

A second case study has been presented, which considers FTC strategies for sensors-faults resilience. Flow-meter and generator rotational speed/position trans-ducers failures have been investigated. Extensive simulations have been carried out on a direct-drive fixed-pitch marine current turbine based on permanent magnets synchronous generators. The FTC strategies have been assessed in terms of energy conversion efficiency, torque ripples, robustness against parameters uncertainty, and computational cost. Regarding, flow-meter sensors fault, simulation results show that turbine parameters-based FTC strategies outperform the other techniques in terms of response time, reduced torque ripples, and energy conversion efficiency. However, their accuracy depends on the turbine parameters knowledge. In the case of rotational speed/position speed fault, EKF-based methods achieve better performance but their implementation in real-world applications is highly constrained by their higher computational burden.

All case studies presented within this chapter could be used as guidelines by industrial and researchers in academia for the choice of the most appropriate fault–tolerant strategy in the marine current turbines context and beyond. Indeed, these approaches could be generalized to all power conversion systems in electrical energy production based on renewables, its conversion, distribution, and consumption. It is worth to mention that the choice of the adapted strategy for fault-resilience is not an easy task and various FTC methods could be appropriate depending the context and the constraints of the energy conversion system.

References

[1]   Melikoglu M. Current status and future of ocean energy sources: a global review. *Ocean Engineering*. 2018;148:563–573.

[2]   Benbouzid MEH, Astolfi JA, Bacha S, *et al*. Concepts, modeling and control of tidal turbines. In: *Marine Renewable Energy Handbook*, pp. 219–278, Wiley, ISTE; Paris, 2011 (Chapter 8), ISBN: 978-1-84821-332-6.

[3]   Titah-Benbouzid H and Benbouzid M. Biofouling issue on marine renewable energy converters: a state of the art review on impacts and prevention. *International Journal on Energy Conversion*. 2017;5(3):67–78.

[4]   Elasha F, Mba D, Togneri M, *et al*. A hybrid prognostic methodology for tidal turbine gearboxes. *Renewable Energy*. 2017;114:1051–1061.

[5]   Touimi K, Benbouzid M, and Tavner P. Tidal stream turbines: with or without a Gearbox? *Ocean Engineering*. 2018;170:74–88.

[6]   Amirat Y, Feld G, Elbouchikhi E, *et al*. Design and applications of a tidal turbine emulator based on a PMSG for remote load. In: *Proceedings of the 2017 IEEE IECON*. Beijing, China: IEEE; October–November 2017, p. 2437–2441.

[7]   Zhou Z, Benbouzid M, Charpentier JF, *et al*. Developments in large marine current turbine technologies—a review. *Renewable and Sustainable Energy Reviews*. 2017;71:852–858.

[8]   Toumi S, Elbouchikhi E, Amirat Y, *et al*. Magnet failure-resilient control of a direct-drive tidal turbine. *Ocean Engineering*. 2019;187:106207.

[9]   Faiz J and Mazaheri-Tehrani E. Demagnetization modeling and fault diagnosing techniques in permanent magnet machines under stationary and nonstationary conditions: an overview. *IEEE Transactions on Industry Applications*. 2017;53(3):2772–2785.

[10] Eriksson S, Bernhoff H, and Leijon M. FEM simulations and experiments of different loading conditions for a 12 kW direct driven PM synchronous generator for wind power. *International Journal of Emerging Electric Power Systems*. 2009;10(1):1–17.

[11] Pham HT, Bourgeot JM, and Benbouzid MEH. Comparative investigations of sensor fault-tolerant control strategies performance for marine current turbine applications. *IEEE Journal of Oceanic Engineering*. 2017;43(4):1024–1036.

[12] Toumi S, Amirat Y, Elbouchikhi E, *et al.* A comparison of fault-tolerant control strategies for a PMSG-based marine current turbine system under generator-side converter faulty conditions. *Journal of Electrical Systems*. 2017;13(3): 472–488.

[13] Pham HT, Bourgeot JM, and Benbouzid M. Fault-tolerant finite control set-model predictive control for marine current turbine applications. *IET Renewable Power Generation*. 2018;12(4):415–421.

[14] Zhou Z, Elghali SB, Benbouzid M, *et al.* Tidal stream turbine control: an active disturbance rejection control approach. *Ocean Engineering*. 2020;202:107190.

[15] Mekri F, Elghali SB, and Benbouzid MEH. Fault-tolerant control performance comparison of three-and five-phase PMSG for marine current turbine applications. *IEEE Transactions on Sustainable Energy*. 2013;4(2): 425–433.

[16] Pham HT, Bourgeot JM, and Benbouzid M. Fault-tolerant finite control set-model predictive control for marine current turbine applications. *IET Renewable Power Generation*. 2018;12(4):415–421.

[17] Amin AA and Hasan KM. A review of fault tolerant control systems: advancements and applications. *Measurement*. 2019;143:58–68.

[18] Hong J, Hyun D, Lee SB, *et al.* Automated monitoring of magnet quality for permanent-magnet synchronous motors at standstill. *IEEE Transactions on Industry Applications*. 2010;46(4):1397–1405.

[19] Kim KC, Kim K, Kim HJ, *et al.* Demagnetization analysis of permanent magnets according to rotor types of interior permanent magnet synchronous motor. *IEEE Transactions on Magnetics*. 2009;45(6):2799–2802.

[20] Xu P, Shi K, Sun Y, *et al.* Analytical model of a dual rotor radial flux wind generator using ferrite magnets. *Energies*. 2016;9(9):672.

[21] Sjökvist S, Eklund P, and Eriksson S. Determining demagnetisation risk for two PM wind power generators with different PM material and identical stators. *IET Electric Power Applications*. 2016;10(7):593–597.

[22] Elghali SEB, Balme R, Le Saux K, *et al.* A simulation model for the evaluation of the electrical power potential harnessed by a marine current turbine. *IEEE Journal of Oceanic Engineering*. 2007;32(4):786–797.

[23] Zhou Z, Scuiller F, Charpentier JF, *et al.* Power smoothing control in a grid-connected marine current turbine system for compensating swell effect. *IEEE Transactions on Sustainable Energy*. 2013;4(3):816–826.

[24] Zhou Z, Scuiller F, Charpentier JF, *et al.* Power control of a nonpitch-able PMSG-based marine current turbine at overrated current speed with flux-weakening strategy. *IEEE Journal of Oceanic Engineering*. 2015;40(3): 536–545.

[25] Kumar D and Chatterjee K. A review of conventional and advanced MPPT algorithms for wind energy systems. *Renewable and Sustainable Energy Reviews*. 2016;55:957–970.

[26] Zafarani M, Goktas T, and Akin B. A comprehensive magnet defect fault analysis of permanent-magnet synchronous motors. *IEEE Transactions on Industry Applications*. 2015;52(2):1331–1339.

[27] Farooq J, Djerdir A, and Miraoui A. Analytical modeling approach to detect magnet defects in permanent-magnet brushless motors. *IEEE Transactions on Magnetics*. 2008;44(12):4599–4604.

[28] Guo B, Huang Y, Peng F, *et al.* General analytical modeling for magnet demagnetization in surface mounted permanent magnet machines. *IEEE Transactions on Industrial Electronics*. 2019;66(8):5830–5838.

[29] Beltran B, Benbouzid MEH, and Ahmed-Ali T. Second-order sliding mode control of a doubly fed induction generator driven wind turbine. *IEEE Transactions on Energy Conversion*. 2012;27(2):261–269.

[30] Levant A and Alelishvili L. Integral high-order sliding modes. *IEEE Transactions on Automatic Control*. 2007;52(7):1278–1282.

[31] Fridman L and Levant A. Higher order sliding modes. In: *Sliding Mode Control in Engineering*, New York, NY: Marcel Dekker, 2002, pp. 53–101 (Chapter 3).

[32] Benelghali S, Benbouzid MEH, Charpentier JF, *et al.* Experimental validation of a marine current turbine simulator: application to a permanent magnet synchronous generator-based system second-order sliding mode control. *IEEE Transactions on Industrial Electronics*. 2011;58(1):118–126.

[33] Qiao W, Zhou W, Aller JM, *et al.* Wind speed estimation based sensorless output maximization control for a wind turbine driving a DFIG. *IEEE Transactions on Power Electronics*. 2008;23(3):1156–1169.

[34] Simoes MG, Bose BK, and Spiegel RJ. Design and performance evaluation of a fuzzy-logic-based variable-speed wind generation system. *IEEE Transactions on Industry Applications*. 1997;33(4):956–965.

[35] Lin TC, Zhu ZQ, and Liu J. Improved rotor position estimation in sensorless-controlled permanent-magnet synchronous machines having asymmetric-EMF with harmonic compensation. *IEEE Transactions on Industrial Electronics*. 2015;62(10):6131–6139.

Chapter 4

Tidal stream turbine monitoring and fault diagnosis

Yassine Amirat[1], Elhoussin Elbouchikhi[1], Claude Delpha[2], Mohamed Benbouzid[3], Demba Diallo[4] and Tianzhen Wang[5]

Condition monitoring of tidal stream turbines is a challenging task of paramount importance due to system location and operation conditions. In this regard, there is a clear need for high reliability, given the severe maintenance access limitations. This chapter reviews fault detection and diagnosis solutions for marine energy conversion system-based tidal stream turbines.

4.1 Introduction

Marine energy has attracted more attention and has become a focal point in recent years due to its reproducibility and cleanness. Although still in the research and development stage and not yet in the commercialization phase, promising marine energies include:

- Wave energy involves converting the energy contained in ocean waves to generate electricity. Converters include oscillating water columns that compress air pockets to drive a turbine. There are also oscillating body converters that use wave motion; and overflow converters that use height differences.
- Tidal energy, produced either by tidal technologies using a dam or impoundment to harvest energy between high and low tide, or by tidal current or tidal flow technologies, or by hybrid applications.
- Salinity gradient energy results from differing salt concentrations when a river empties into an ocean.

[1]ISEN Yncrea Ouest, L@bISEN, France
[2]University of Paris-Saclay, CentraleSupelec, CNRS, Laboratoire des Signaux et Systèmes, France
[3]University of Brest, CNRS, Institut de Recherche Dupuy de Lôme, France
[4]Université Paris-Saclay, CentraleSupelec, CNRS, Group of Electrical Engineering Paris, France
[5]Shanghai Maritime University, Logistics Engineering College, China

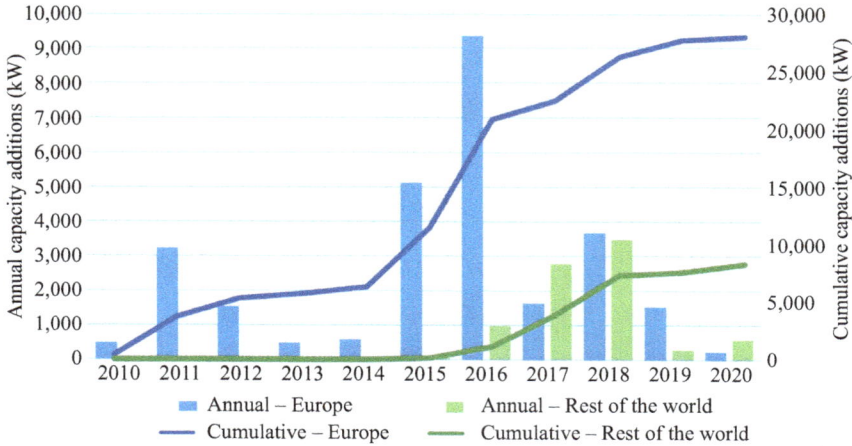

Figure 4.1   Installed global tidal stream energy capacity [1]

- Ocean thermal energy conversion generates power from the temperature difference between warm surface seawater and cold seawater at 800–1,000 m depth.

Figure 4.1 illustrates the annual and cumulative installed tidal turbines in Europe and the rest of the world. It clearly shows a slowdown in cumulative wave energy installations in recent years. This can be explained by the fact that there is not a lot of feedback to allow for the massive deployment of these turbines. Additionally, it should be noted that many projects are currently in the manufacturing phase onshore. Moreover, the operational problems encountered worldwide during the 2020 pandemic have induced a negative impact on this deployment.

Tidal stream turbines can make a significant contribution to the supply of renewable energy production. Unfortunately, the potential for utility-scale deployments will depend not only on the economics of power generation but also on the likely hydro-environmental effects of deployments. These may include physical, acoustic, chemical, and electromagnetic effects, but the effects of energy removal are considered the most significant [2]. Hence, operating and maintaining a utility-scale tidal turbine in such environment is particularly challenging, and data collection and analysis play a vital role in doing so successfully.

TST energy devices have moved beyond the prototyping phase. They are beginning to be installed at different scales of power, from a few kW (as an example, the Guinard energies tidal turbine P60 and P154 from [3]) to a few MW [3].

4.1.1   Tidal stream turbine challenges

So far, the most widely used converters for harvesting energy from marine currents are tidal stream turbines. Different architectures have been proposed; two main TSTs have been identified: horizontal axis as well as vertical axis TSTs. However, the horizontal

(a)

(b)

(c)

*Figure 4.2   Horizontal axis tidal steam turbines. (a) Guinard Energies P66 tidal
stream turbine. (b) Oceade tidal stream turbine [4]. (c) AR1000 tidal
stream turbine [5].*

axis architecture is likely to become the industry standard. Figure 4.2 illustrates some
of them.

Tidal stream turbines, unlike many other renewable energy sources, are a reliable
form of kinetic energy produced by regular and periodic tidal cycles. Unlike wind
power, energy production from tidal currents is not influenced by weather conditions

Table 4.1   Comparison between WEC and TST parameters for the
            same capacity [6]

Item	WEC	TST
Density	1.22 kg/m^3	1,000 kg/m^3
Velocity	12 m/s	2.6 m/s
Rotor diameter	27.1 m	9.31 m
Thrust load	146 kN	675 kN
Torque	546 kNm	837 kNm

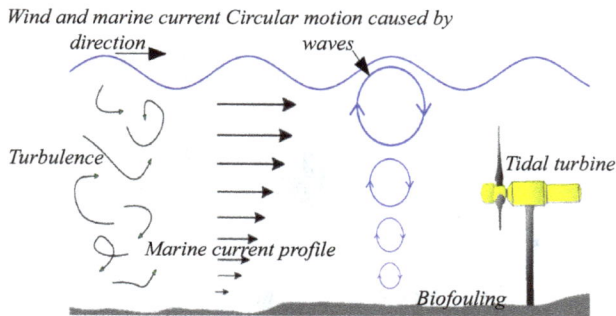

Figure 4.3   *Tidal stream turbine harsh environment conditions*

and is predictable hundreds of years in advance. Another advantage of harnessing tidal stream energy is that the density of water is very high, which explains why the blades can be smaller and rotate more slowly while producing a significant amount of energy. Table 4.1 shows a comparison between wind energy converters (WECs) and TST for the same capacity.

In addition, this natural predictability of tidal power is very interesting for grid management, as it means that we are no longer dependent on fossil fuel power stations. TSTs are mounted on the seabed at sites with high tidal and continuous ocean current speeds, where they capture the energy of the flowing water.

Currently, tidal power technology has yet to prove its long-term operational avail-ability and reliability. The harsh marine environments and accessibility are indexed as the main issues for maintenance, and this can accentuate issues of availability and reliability. Thus, minimizing the uncertainty surrounding the operation and main-tenance of these devices will therefore be crucial to achieving economically viable energy extraction. Figure 4.3 depicts the severe environment in which TSTs operate. It shows that TSTs are subject to random current waves and high turbulence effects from the irregular seabed. TSTs are designed to operate in high and variable marine current velocity sites where a considerable amount of kinetic energy can be trans-formed into electricity. This clearly highlights that TSTs will be subjected to very challenging conditions and constraints throughout their service life.

Figure 4.4 Contamination of a TST Guinard P154. (a) Contamination of the base. (b) Contamination of the fairing.

The environmental constraints make tidal stream turbines potentially suffer from a higher failure rate. These failures are mainly related to mechanical subsystems and electrical subsystems [7]. The most indexed causes of failure are solid particle contamination, insufficient lubricant, high mechanical load, corrosive contamination, cyclic mechanical load, impact loads and vibration loading. For illustration, Figures 4.4 and 4.5 present organic contamination of the base and the fairing of a tidal stream turbine prototype after 4 and 6 months of immersion, respectively. Consequently, condition monitoring and fault diagnosis are essential to enhance energy extraction from marine current turbines.

4.2 Condition monitoring and fault diagnosis in TST

In condition monitoring systems, there is no predictive maintenance without a minimum diagnosis of defects and their severity estimation. For these reasons, the first step of monitoring action is to ask which defects are likely to occur on the machine

Figure 4.5   Attachment of marine current turbine after 6 months of launching

to be monitored. The second step is to find out how these faults manifest themselves. What information and what descriptive parameters of the defect should be? Furthermore, measuring to have the correct information will allow stating whether the situation is healthy or not (anomaly detection), but also that will enable finding the origin and (isolation of the source and severity of the anomaly). For this purpose, deep knowledge about all the phenomena involved during a fault constitutes an essential background for developing any fault diagnostic system. To that end, a failure mode analysis must be carried out.

4.2.1   Failure modes

To develop efficient fault detection and diagnosis system, it is necessary to have a thorough knowledge of all the phenomena involved in the occurrence of a fault. Suppose we consider a fault as a particular input acting on the TST. In that case, a diagnostic system must detect its event and isolate it from all other information (operating references and sensors feedbacks) and environmental disturbances. For the fault detection problem, we would like to know whether or not a fault exists in the power conversion system through the analysis and processing of available or estimated measurements. This approach has been widely used in several domains and fields and provided enough information on the history of focusing on healthy or faulty modes. However, the limited availability of failure data related to marine turbine implementations has prompted the research community to develop approaches to fault modes analysis based on the analysis of a failure modes and effects analysis (FMEA) constructed from data applicable in similar industries, particularly in the wind energy field, or by developing generic numerical models as in [8]. A quantitative analysis of actual wind turbine failure data has shown important features of failure rate values and trends. For illustration, Figure 4.6 checks off the failure number distribution for

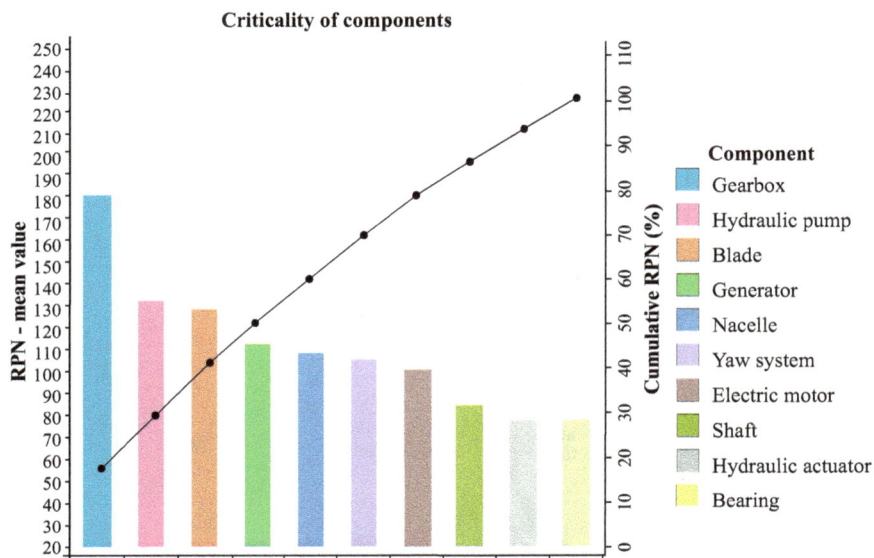

Figure 4.6   Criticality of tidal turbine components [9]

the main components of a wind turbine conversion system. This approach can serve as a base for assessing failure modes analysis of tidal stream turbines.

After the criticality study, the most critical assemblies and sub-assemblies for a wind turbine or generic TST models are summarized in Figure 4.7 and at the sub-assembly level is enumerated in Table 4.2.

Table 4.2 shows a summary of the most critical assemblies and sub-assemblies found during the FMEA. These critical assemblies must be integrated into the condition monitoring system (CMS). Monitoring and condition-based maintenance (CBM) strategies are based on the relevant components. In contrast to wind turbines, the yaw mechanism is not critical for tidal turbines, as it is only used to reverse the assembly at slack tide, operating at a low load. As a result, most failures were linked to the gearbox, followed by the hydraulic system and blade/pitch components.

4.2.2   Condition monitoring

Condition monitoring of tidal stream turbines is a broad scientific area. Its ultimate purpose is to ensure this complex electro-mechanical energy conversion system's safe, reliable, and continuous operation. Many techniques and tools are available to monitor the electro-mechanical systems, thus prolonging their life span. Some of the technologies used for monitoring include sensors, which may measure speed, output torque, vibrations, temperature, flux densities, etc. These sensors are managed together with algorithms and architectures, which allows efficient monitoring of the system. The most popular methods of rotating system condition monitoring utilize

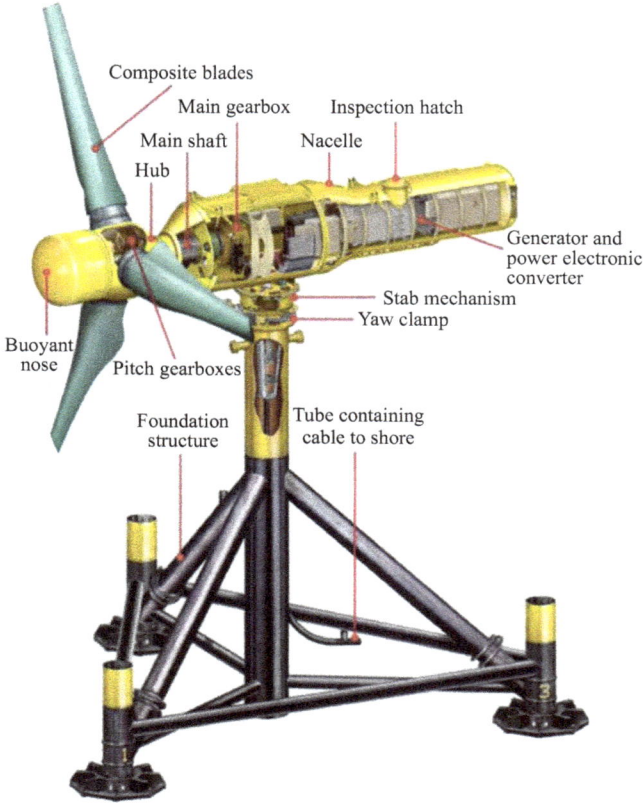

Figure 4.7   Generic tidal turbine of the "complex" RealTide concept category showing sub-assemblies discussed in the following [10]

Table 4.2   Most critical assemblies during the FMEA

Critical assembly	Critical sub-assembly
Electrical system	Power electronic converter, generator, transformer(s)
Drivetrain	Low speed shaft, gearbox
Rotor	Blades, pitch system

the steady-state spectral components of the mechanical or electrical quantities. These spectral components include vibration, rotating speed or voltage, current and electric power. They are used to detect gearbox failures, bearing failures, unbalanced defects or turn faults, broken rotor bars and air gap eccentricities, and so on. Different electric machine conditions monitoring and fault diagnosis procedures are proposed in the literature of the electrical engineering community. The well-established methods, from the theoretical and experimental point of view, are motor current signature analysis (MCSA), extended Park vector approach (EPVA), instantaneous electric power spectral analysis (IEPSA), and vibration monitoring. However, regardless of the monitoring method or technique, it is necessary to use defect indicators. These indicators are physical quantities used regularly to monitor the installations. Their evolution allows them to alert of the degradation, and we can distinguish two faults' indicators:

- scalar indicators or global levels;
- shape indicators or spectra.

These indicators give the machine's signature during the first campaign of measurements on the machine in healthy operating condition (or supposed to be).

4.2.2.1 Vibration monitoring (VM)

TST has complex electro-mechanical structures that oscillate, and the coupled parts transmit these oscillations. Hence, the operation of the TST generates forces that often cause subsequent failures (rotating forces, turbulence, shocks, and instability). These forces cause vibrations that will damage the structures and components of the system. Analyzing these vibrations will make it possible to identify the forces as soon as they appear before they cause irreversible damage. It will also allow, after analysis, to deduce the origin and estimate the risk of failure. The condition-based maintenance is based on these concepts. So, to implement CBM, it is necessary to determine the most frequent causes of failure, evaluate their costs and probability of occurrence, and set up a policy to detect the symptoms as early as possible. Moreover, mechanical failures have been a persistent problem in many industries, accounting for a significant proportion of all failures in rotating machines. For example, bearing failure of an electric drive or rotating electric generation system is the most common failure mode associated with extended downtime. Bearing failure is typically caused by improper lubrication, occasionally manufacturing faults in the bearing components, and some misalignment in the drive train, leading to abnormal loading and accelerating bearing wear. Therefore, the basis of vibration-based fault detection is to detect the increase in the signal's energy, or rather, if we want to have early detection, of what in the energy indicates the presence of repeated small shocks. The techniques are multiple and consist in analyzing the signal provided by an accelerometer. The installation of the sensors on the machines is also essential, and the placement of the accelerometers determines their sensitivity to increased vibrations [11]; hence, each measurement campaign must be carried out at precise points and always the same ones. Indeed, a mechanical phenomenon can give sensitively different vibratory images depending on the point of measurement. For the assembly shown in Figure 4.8, the motor has two bearings noted M1 and M2; the pump also has two bearings noted P1 and P2, and each bearing is monitored in three orthogonal directions vertical (V), horizontal (H) and axial (A). After, we have a certain number of descriptors presented in Table 4.3, such as RMS value of acceleration, peak value or peak to peak, crest factor, or a combination of these values. Other more complex quantities can be added to these fundamental quantities, considered by some authors as more representative of this type of defect. One is the Kurtosis: this quantity is close to a form factor. We can also be interested in

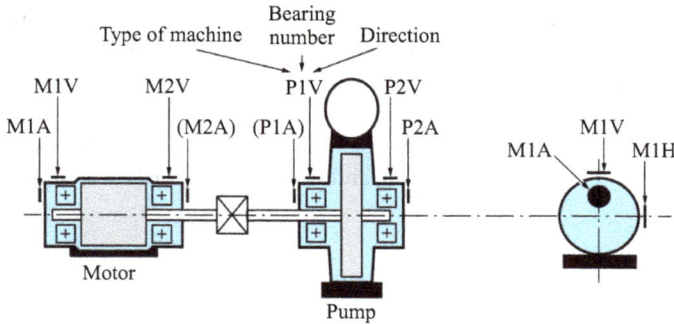

Figure 4.8　Measuring points and vibration sensors' placement scheme

Table 4.3　Descriptors in vibration monitoring

RMS value of the acceleration	δ_{rms}
Pick value	δ_p
Pick to pick value	δ_{pp}
Mixed value	$\dfrac{\gamma_{pp}}{\gamma_{rms} + \epsilon\gamma_{rms}}$

the spectrum of the envelope of the vibratory signal if we want to locate the defect. It is calculated from the demodulated and filtered signal and contains the characteristic frequencies of the faults. However, the more complex the transmission system is, the more the number of its accelerometers increases; for example, monitoring the gearbox of a wind turbine, as shown in Figure 4.9, will require the use of six accelerometers. It should be noted that the accelerometer technology also depends on the type of fault to be detected.

In the criticality analysis and the experience feedback, bearings are indexed as the most stressed components of machines and are a frequent source of failure. The defects that can occur are as follows: spalling, seizure, corrosion, pitting, etc. All these defects have one thing in common: sooner or later, they result in the loss of metal fragments. This precursor defect to destruction is spalling. It results in repeated impacts of the balls on the bearing cage. A plethora of research works [13,14] states that due to the construction of rolling-element bearings, a defect generates a precisely identifiable signature on vibration, and the generated frequencies present a practical path for monitoring progressive bearing degradation. Spectral analysis of vibrations has been used in rotating machines' fault diagnosis for decades. It has been traditionally used for monitoring bearing and rotor faults. For instance, Figure 4.10 gives an example of the vibration spectrum of a rotating machine. Indeed, the characteristic frequencies,

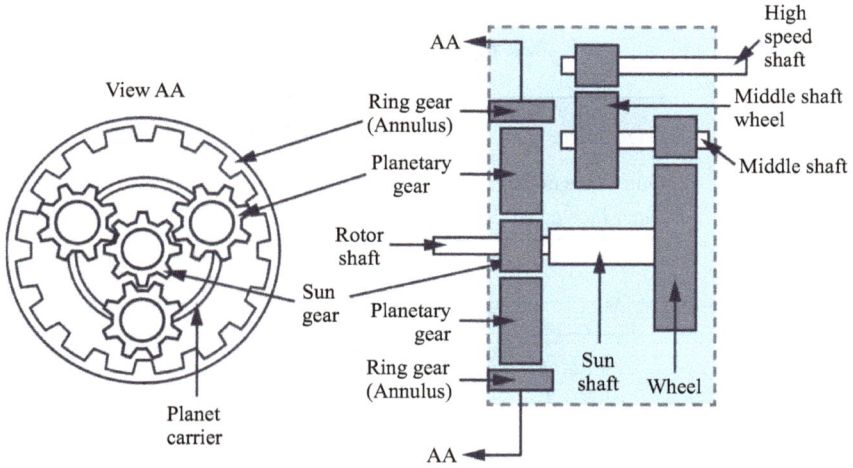

Figure 4.9 Gearbox structure [12]

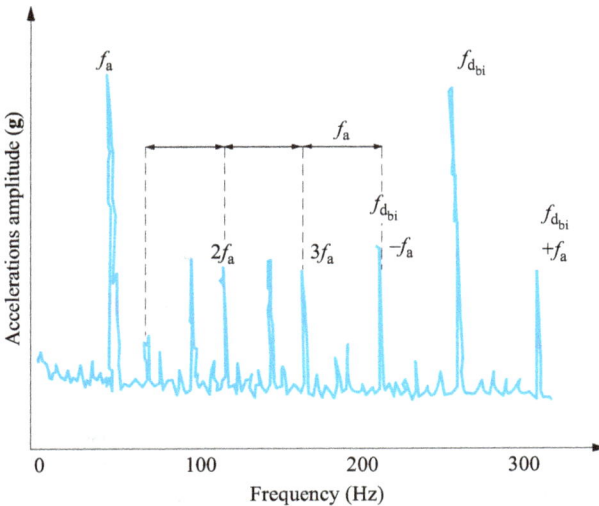

Figure 4.10 Vibration spectrum example

depicted in Figure 4.10, of the defects located on parts of a bearing are expressed in Table 4.4.

Many devices allow an excellent detection of bearing anomalies. Their goal is to detect repeated shocks as early as possible. However, in the early stages of spalling, the shock, which is of short duration, does not modify the average energy of the system. Therefore, it is not seen if the vibration level is studied. So, filtering by

Table 4.4   Characteristic frequencies of
bearing defects

Cage defect	$f_{d_{ca}} = \frac{1}{2}f_a(1 - (d/D))\cos\phi$
Outer race defect	$f_{d_{be}} = Nf_{d_{ca}}$
Inner race defect	$f_{d_{bi}} = N(f_a - f_{d_{ca}})$

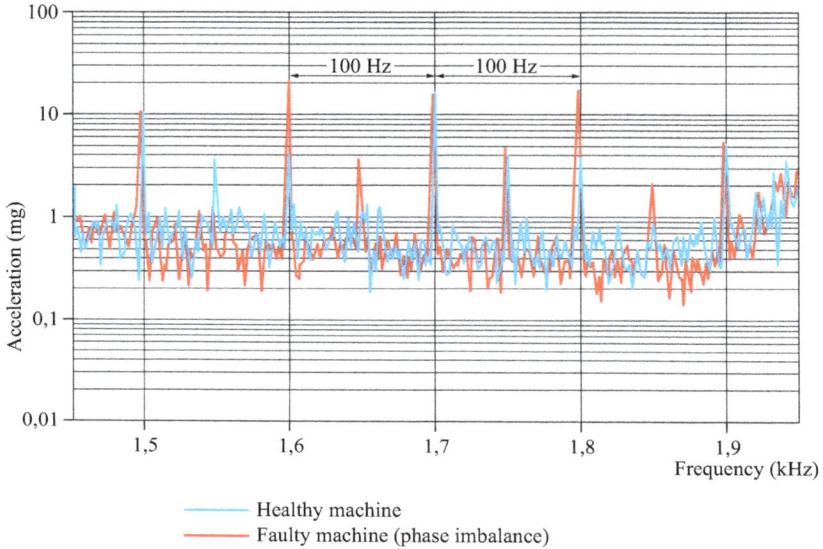

Healthy machine
Faulty machine (phase imbalance)

Figure 4.11   Zoom FFT [1,450–2,050 Hz] centered around 1,700 Hz (notch frequency)

the vibration sensor is performed to improve detection. For this purpose, a wide bandwidth accelerometer with a low-damped resonance is used, which, excited by shocks, will respond to its resonance and act as a selective filter. It lets the shocks through, not the background noise. This filtering provides better discrimination of the defect. It suffices to measure the output signal level, which, in the absence of a shock, is low and increases very quickly in the event of a fault. Therefore, this type of fault is characterized by an increase in the signal's efficacy and crest factor. The vibration analysis is also investigated to detect electrical faults. For illustration, Figure 4.11 depicts the zoom FFT of the vibration signal for the healthy and faulty machine, where the fault is the phase imbalance. However, experience and industrial feedback have demonstrated that vibration monitoring has made out its efficiency, and it is highly suitable for rolling-element bearings, shaft misalignment, and eccentricity; it represents an issue when requiring a good vibration baseline [15]. If no baseline is available, no history has been built, making detecting the specific frequencies impossible when the background noise has risen [14]. To overcome this issue, many alternatives have emerged in the electro-mechanical community by analyzing the electrical quantities, particularly the stator current of the electrical machines for both motoring and generating modes. These alternatives are known as machine current signature analysis (MCSA), including the use of electrical current [15,16], or the instantaneous power factor [17].

4.2.2.2 MCSA

Most mechanical or electrical faults can manifest their impact in electric quantities; some can generate additional frequencies, and others cause amplitude and or phase modulation. Consequently, electric quantities signature analysis is considered a path for condition monitoring. This technique is now widely used to detect and diagnose faults in electric machines operating in motoring or generating modes, and the most popular is the machine current signature analysis. For illustration, let us assume that in a healthy condition, the supply current is expressed as:

$$i_{sA} = i_M \cos(\omega t - \alpha)$$
$$i_{sB} = i_M \cos(\omega t - \alpha - 2\pi/3), \tag{4.1}$$
$$i_{sC} = i_M \cos(\omega t - \alpha + 2\pi/3)$$

In a faulty state, for example, for a broken rotor bar of an induction machine, the line current can be expressed as follows:

$$i_{sA}(t) = i_F \cos(\omega t - \alpha) + i_{dL} \cos[(1 - 2s)\omega t - \beta_L]$$
$$+ i_{dR} \cos[(1 + 2s)\omega t - \beta_R] \tag{4.2}$$

We note the appearance of sideband frequency components around the fundamental component as depicted in Figure 4.12. Analysis of location and amplitude of the sideband frequency components is generally used to detect broken rotor bars, shorted turns and abnormal levels of air-gap eccentricity. Lower sideband frequency components are specifically due to broken rotor bars, while the upper sideband frequency components are due to consequent speed oscillations. The advantage of this method is the availability of stator current that is usually measured for motor protection and control. The disadvantage is that the sideband frequency components usually differ depending on the fault types and operating conditions. The electric machine load inertia also affects the magnitude of these sidebands [18,19].

For current spectral estimation based on the fast Fourier transform (FFT) and its extension, the short-time Fourier transforms (STFTs) have been widely employed [20]. Due to these techniques frequency resolution limitation [21], high-resolution technique: MUltiple SIgnal Classification (MUSIC) [22] and ESPRIT [23,24] were afterwards investigated. However, these techniques have several drawbacks since they are difficult to interpret, and extracting variations in the time domain for non-stationary signals is challenging. To overcome this problem, procedures based on time–frequency representations (Spectrogram, Quadratic representation such as Wigner Ville, etc.) [25–27], or time-scale analysis (wavelet

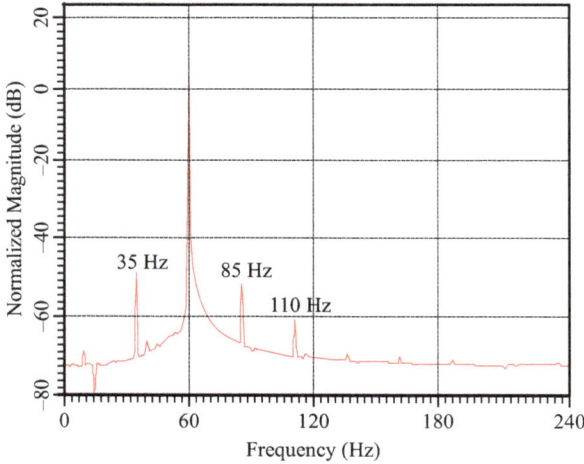

Figure 4.12   Stator current spectrum

transform) have been proposed in the literature of electric machines community [28–30]. There are also parametric methods based on parameter estimation of a known model [21]. Nevertheless, these methods are formulated through integral transforms and analytic signal representations [31], so their accuracy depends on data length, stationarity, and model accuracy. Most electric machine faults lead to current modulation (amplitude and/or phase) [32]. This is the particular case of bearing faults [33]. Indeed, a bearing fault is assumed to produce an air-gap eccentricity [26] and, consequently, an unbalanced magnetic pull. Hence, this gives rise to torque oscillations, which lead to amplitude and/or phase modulation of the stator current [15,26,34].

4.2.2.3   Instantaneous electric power analysis (IEPA)

Besides the generator current, the voltage is easily accessible and can give, in combination with the current, the extra information that is the instantaneous electric power necessary to analyze the condition of the electric generator. The partial instantaneous electric power is given by:

$$P(t) = \frac{3}{2} u_M i_M \cos(\phi) \tag{4.3}$$

where $\cos(\phi)$ is the power factor.

In a faulty condition, we note the appearance of the sideband frequency components in the total instantaneous electric power, for example, that can be expressed as [35]

$$P(t) = \frac{3}{2} U_M \left[i_F \cos(\phi) + i_{dL} \cos(2\omega t + \beta_L) + i_{dR} \cos(2s\omega t - \beta_R) \right] \tag{4.4}$$

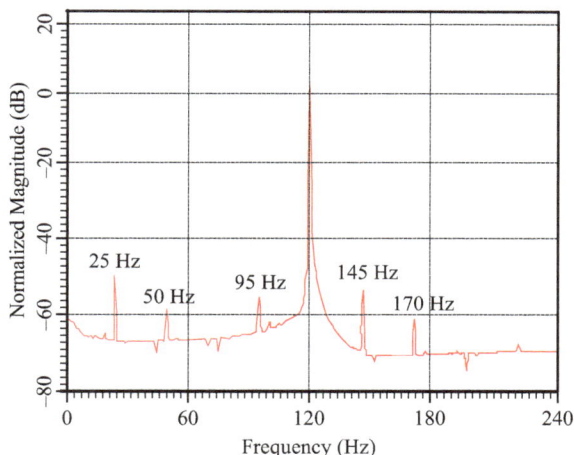

Figure 4.13   Instantaneous electric power spectrum

The partial or total instantaneous electric power spectral analysis (IEPSA) is one of the most used methods [35]. Figure 4.13 gives an example of the instantaneous electric power spectrum.

4.2.2.4   Model-based or data-driven approaches

In addition to spectrum processing and analyzing approaches of electrical or mechanical quantities, fault detection and diagnosis of electrical systems can be carried out using model-based or data-based approaches [36]. While model-based techniques require the use of almost accurate models including the machine parameters, signal-based approaches have the advantage of being driven by data without any prior knowledge about the monitored system and the need for the acquisition of the signals reflecting the state of health [37]. Then, useful information is extracted from noisy signals, and most of the time statistical proprieties are investigated for failure detection and diagnosis [38]. However, to automate the diagnosis process and minimize human involvement [39,40], the signal processing techniques are often combined with artificial intelligence tools [41]. In terms of artificial intelligence, machine learning seems to be the solution of choice to effectively address major issues faced by data-driven failure detection and diagnosis approaches [42]. The advent of industry 4.0 has encouraged the use and investigation of artificial intelligence approaches for fault detection and diagnosis in various industrial fields, and a plethora of literature addressed this topic. In this context, the approach of convolution neural networks (CNN) has been indexed to be well adapted for features extraction [43,44], while recurrent neural network (RNN), with its long short-term memory (LSTM) new variant, better suited for learning and classifying time series [45,46].

For instance, in [47], it suggests an approach illustrated in Figure 4.14, for the early diagnosis of bearing faults that are based on the variational mode decomposition

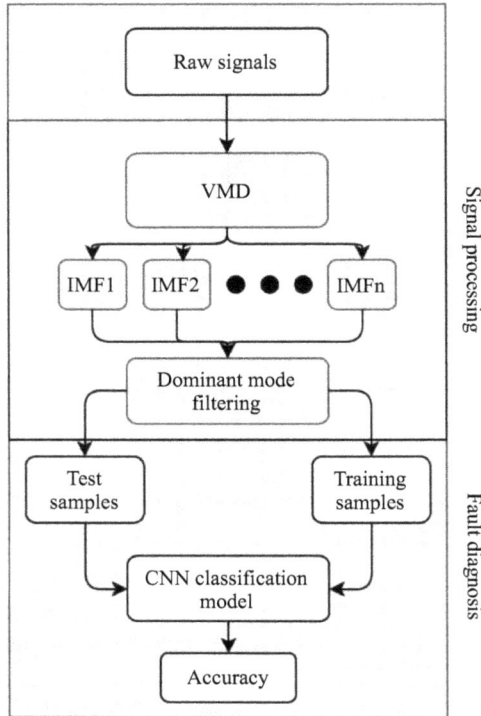

*Figure 4.14   Artificial intelligence-based failure diagnosis methodology
flowchart [47]*

(VMD), which is a data-driven approach, used as a notch filter for dominant mode cancellation. Additionally, the author suggests a machine learning method, specifically the one-dimensional convolution neural network (1D-CNN), for the purposes of detection and diagnosis. To be more specific, the method proposes first extracting features with VMD for fault detection, and then transitioning to multi-scale feature extraction with CNN convolution and pooling layers for classification and diagnosis.

4.2.2.5   Image processing based-fault detection

In recent years, fault detection and diagnosis-based image processing has become a focal point in the research community and indexed as a powerful tool for condition monitoring, as it can provide non-contact measurements with good accuracy. It can be used to detect changes in the texture or color of a surface, such as corrosion, or to detect cracks in structure or debris [48], to detect anomalies in industrial production processes, and can also be used to detect changes in electrical parameters in electric power generation system, such as in photovoltaic systems [49] and other features of a system allowing continuous monitoring of its condition. Additionally, image processing can provide insights into the condition of an object that may not be visible to the human eye, allowing for more comprehensive monitoring. With the improvement of computer technology and mathematical theory, there are now more and better ways to analyze images and extract features by investigating machine learning and deep learning tools [50,51]. For traditional machine learning, complex feature extractors must be made for specific cases to retrieve the desired features. When compared to traditional machine learning, the main benefit of deep learning is that these rich features are not made by human engineers. Instead, CNNs learn them automatically from raw data [52], and deep learning has been shown to be very good at finding complex structures in data with a lot of dimensions [53]. So, for condition-based monitoring systems that use image processing to find defects, deep learning can be a big part of the start of the era of intelligent detection with machine vision. Based on machine vision, there are three main types of tasks that can be used to find defects in industrial production: classification, localization, and segmentation. Some simple image preprocessing methods can help with the image analysis that comes next, and they can also be used to find a few simple defects. In most cases of defect detection, you need more image processing methods to get enough features to understand what the defect is. The most common types of deep learning network architecture for learning image features are CNNs, deep belief networks (DBNs), and stacked auto-encoders (SAEs). Also, long-term short-term memory (LSTM) [54] is a very important part of how we remember images that change over time. DBNs and SAEs can help improve the effectiveness and accuracy of multi-feature fusion detection. Figure 4.15 depicts a typical architecture of the CNN-based image processing model. Image preprocessing helps the machine better understand the image and prepare for the step of image analysis [55]. Image preprocessing aims to get rid of information that is not useful and get back the information that is useful. Image noise can be caused by various factors, such as the machine vision field environment, the photoelectric conversion of the charge-coupled device (CCD) image sensor, the transmission circuit, and electronic parts. The quality of the image is reduced due to these noises, which has a direct effect on the analysis of the image. So, the image preprocessing step is very important to remove noise from an image and get rid of the noise. In general, spatial domain methods and frequency domain methods are used in the image preprocessing step. The most popular preprocessing algorithms are gray-scale transformation, histogram equalization, and various filtering algorithms based on spatial and frequency domains [56] or mathematical morphology [57]. For instance, the Fourier transform was indexed as the common way to change from the spatial domain to the frequency domain, and its inverse transform can transform the image from the frequency domain back to the spatial domain. In the image processing field, the two-dimensional discrete Fourier transform (DFT) and the inverse two-dimensional discrete Fourier transform (IDFT) are represented as [58]:

$$F(u, v) = \sum_{x=0}^{M-1} \sum_{y=0}^{N-1} f(x, y) e^{-j2\pi(ux/M + vy/N)} , \qquad (4.5)$$

$$f(x, y) = \frac{1}{MN} \sum_{u=0}^{M-1} \sum_{v=0}^{N-1} F(u, v) e^{j2\pi(ux/M + vy/N)} , \qquad (4.6)$$

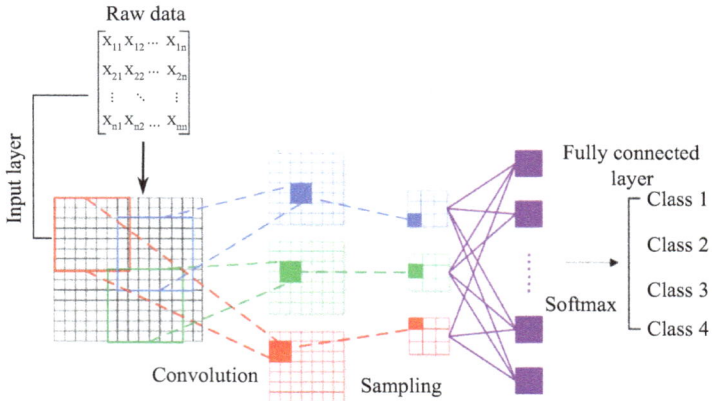

Figure 4.15   Architecture of a CNN model for image processing [58]

where $f(x, y)$ represents a digital image of size $M \times N$, and then the frequency domain representation $F(u, v)$ can be obtained by using DFT formula mentioned above. In formulas (4.1) and (4.2), u ($u = 0, 1, 2, ..., M - 1$) and v ($v = 0, 1, 2, ..., N - 1$) represent the frequency domain variables; $x(x0, 1, 2, ..., M - 1)$ and $y(y = 0, 1, 2, ..., N - 1)$ represent the space domain variables. The data in the frequency domain represent the intensity of gray-scale changes in the image. The frequency domain filtering adjusts the image's Fourier transform before calculating its inverse transform to produce the treated image. For instance, the moving average window filter and Wiener linear filter use a low-pass filter to denoise based on the concept that noise energy is concentrated at high frequency and the picture spectrum is dispersed across a restricted range [59].

4.3   Fault detection and diagnosis in TST

For the TST context, several technological issues, among others, fault detection of TST blades, still require further progress for their successful implementation [60,61]. Indeed, power generation is likely to progressively degrade due to blade faults and then causes disruptive disturbance when the marine current generator is connected to the grid. Fault detection of MCT blades still present several challenges due to the complexity of the underwater environment [2]. Attachments like plankton or biofouling may considerably influence the turbine blade as these may trigger different imbalance faults. In [62], authors studied this issue in the generalized context of marine renewable energy converters, highlighting the biofouling impact and solutions to prevent it, but the analysis of this phenomenon is complex because the biofouling development is a slow process, as illustrated in Figure 4.16.

In literature, various sensors-based methods were proposed to detect the rotor imbalance of TST [63,64]. However, fault detection using stator current signals has many advantages because it costs less and does not need to consider sensors faults

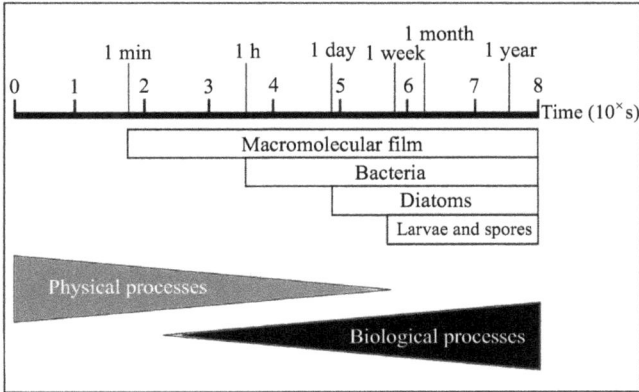

Figure 4.16 Schematic colonization process leading to the establishment of a fouling whole [62]

[65]. The ideal case in various electro-mechanical applications is that the electrical current presents a pure sine wave shape with only the fundamental supply component. However, due to various phenomena such as eccentricity, slots, and saturation, the electric current contains more than the fundamental supply component but also additional harmonics [66], and eventually a component introduced by the failure. In addition, the operating environment of TSTs causes several phenomena, such as turbulence and waves, to interfere with this operation, and the current can be expressed as [67]:

$$x(t) = \sum_{k=1}^{M} a_k(t) \sin(\phi_k(t)). \tag{4.7}$$

Several approaches are proposed to reduce the influence of these interferences and analyse non-stationary and nonlinear signals. The data-driven and iterative algorithm known as empirical mode decomposition (EMD) or its improved version (EEMD), proposed in [68,69], has become a recognized tool in a wide range of applications in signal processing [70] and fault detection in rotating machines [71,72]. For instance, to decompose the stator current signal into different intrinsic mode functions (IMFs), as depicted in Figure 4.17. To illustrate the data-driven decomposition concept, let us assume the synthesized signal $x_{syn}(t)$ given by:

$$x_{syn}(t) = a_1 \sin(\omega_1 t) + a_2 \sin(\omega_2 t), \tag{4.8}$$

where a_1 and a_2 are the amplitudes of the first and the second component, respectively; while ω_1 and ω_2 are pulsations of those components. Figure 4.17 illustrates the result of decomposing $x_{syn}(t)$ through the EEMD algorithm.

In the TST fault detection context, the EMD was investigated to extract the IMF containing the fault signature to detect the imbalance fault [73–75]. In [76,77], the moving average filter method filters the interference in the signal. All these

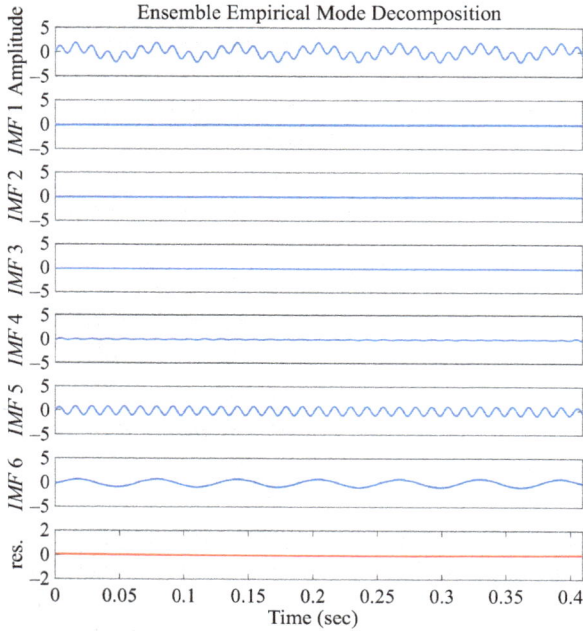

Figure 4.17   EEMD synthetic signal [67]

methods denoise the stator current signal measured from the MCT system and the imbalance fault detection. However, for the strategies using EMD, there are still some problems, such as the end effect and the modal confusion [78]; for techniques using the moving average filter method, only the high-frequency signals can be filtered. As reported in [79], the wavelet transform method is used to preprocess the raw signal, reduce the interference components, and extract valuable information to detect the fault. However, it is difficult for these methods to use the wavelet transform to set the appropriate parameters in practical application. To decrease the data dimensions and automatically detect the fault, the principle component analysis (PCA) is used in [80,81]. This method can build a standard model in a healthy state, and the fault can be detected automatically. However, the PCA method could not be directly applied to the imbalance fault detection because data preprocessing is required.

4.4   Tidal stream turbine faults

The use of marine energy for power generation is growing rapidly. However, the immersion of equipment exposes it to the effects of the marine environment, for example, corrosion, pollution, and fouling by animals and plants. In addition, once installed, the accessibility of the equipment is more challenging to consider inter- ventions in the case of failure. Predictive maintenance is one way to ensure regular monitoring of the components of the generator system: the blades, the generator, the connectors, and possibly the static converter. Thus, early detection and diagnosis of faults increase the availability and reliability of power generation. However, several technological issues among others fault detection of TST blades still require further progress for their successful implementation [60,61]. Indeed, power generation is likely to progressively degrade due to blade faults and then causes disruptive disturbance when the marine current generator is connected to the grid. Fault detection of MCT blades still present several challenges due to the complexity of the underwater environment [2]. It appears that attachment like plankton or biofouling may have a considerable influence on the turbine blade as these may be triggering different imbalance faults [62]. The survey in [61] reviews different blade fault types under wave and turbulence; and index trends in tidal current blade fault detection methods, including multi-domain approaches. This survey shows that built-in sensor-based fault detection methods, which use phase currents and voltages across the generator windings, provide several advantages for MCT blade fault detection.

4.4.1   Impact of imbalance fault

The imbalance faults will affect the operation of the system, which not only degrades the performance of TSTs, but may also significantly damage its structure [82]. In addition, the MCT installed underwater is affected by many factors such as attachments, surges, and turbulence. Moreover, it is difficult to extract the imbalance fault signature. In real applications, the marine environment's variation is uncertain and affecting the stator current signal. The uncertainties caused by the marine environment will cover the valuable information in the stator current signal [83]. Hence, detecting the current signal in the time domain is a challenging task [61].

When an imbalance fault occurs on a blade, its mass distribution will differ from the others. An equivalent imbalance mass will occur and induce a vibration in the shaft rotating speed. As shown in Figure 4.18, m and r_u denote the imbalance mass and the distance to the shaft, respectively. From a mechanical point of view, the

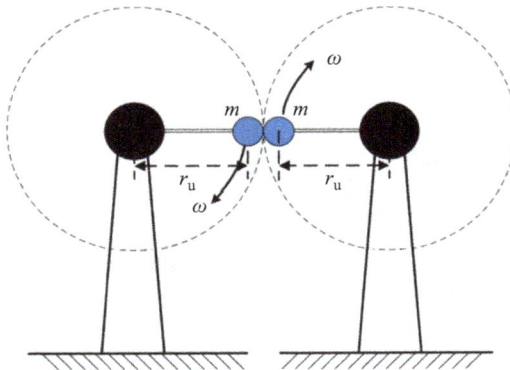

Figure 4.18   The effect of blade imbalance in an MCT [84]

parameters m and r_u will have an impact on the revolution of the shaft of the turbine and consequently can affect the torque on the shaft, and the torque distortion caused by a bigger m and a smaller r_u may be equal to that caused by a smaller m and a bigger r_u. The fault information in the shaft rotation frequency is then transferred to the instantaneous frequency of the stator current signal as expressed in (4.10). Therefore, it is difficult to find the fault information in the time domain.

4.4.2   Current signal model

Under rotor asymmetry, fault frequencies are determined by $f_f = f_s \pm kf_r = (1 \pm k/p)f_s$, where f_r is the rotational speed frequency, f_s is the supply frequency component, p is the pole pairs number, and k is an integer. With the blade imbalance fault, a simplified model, which describes the TST electric generator stator current, can be defined by [85]:

$$i_s(t) = A_t \cos{(p\omega_m t + p\Delta\omega_m t + \varphi)} \tag{4.9}$$

where A_t is the amplitude of the stator current, φ is the initial angle, ω_m is the mechanical rotating speed, and $\Delta\omega_m$ is the rotation speed variations.

By assuming that $\psi = p(\omega_m + \Delta\omega_m)$, the stator current can be expressed as:

$$i_s(t) = A_t \cos{(\psi t + \varphi)} \tag{4.10}$$

4.4.3   MSCA-based imbalance fault detection

To demonstrate this hypothesis, the test bench depicted in Figure 4.19, featuring a direct-drive PMSG prototype with 0.23 kW capacity and eight pole pairs [61,84] the TST operates in a circulating flume, which uses a pump motor to generate controllable flow. The sampling frequency of the data acquisition system is 1 kHz. The blade imbalance fault is emulated by attaching winding ropes to the blade. With the experimental platform, a TST working in the underwater environment, which is affected

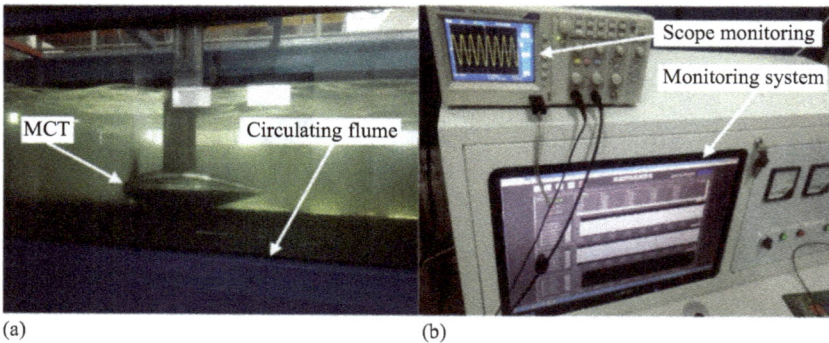

(a)   (b)

Figure 4.19   Test rig for imbalance blade detection [61]. (a) The MCT in the circulating flume. (b) The monitoring system of the MCT.

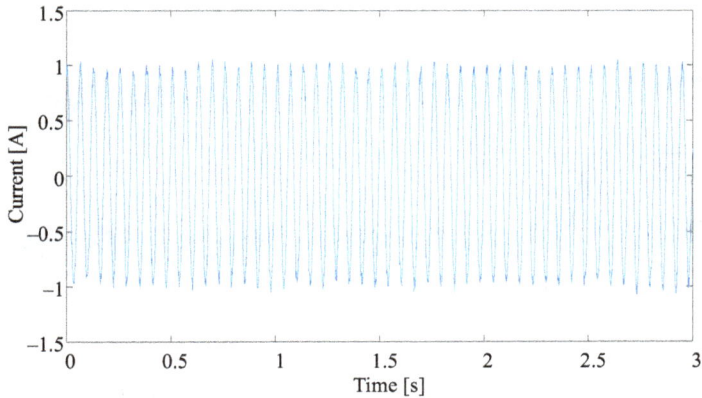

(a)

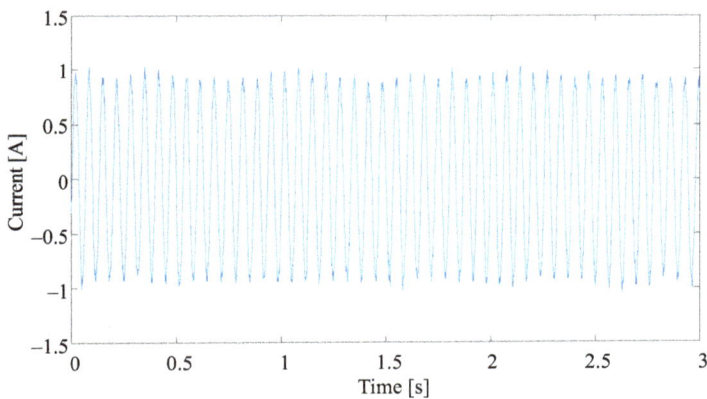

(b)

Figure 4.20 *The stator current signals of the TST in different health states. (a)*
Healthy rotor blades. (b) Imbalanced rotor blades.

by the turbulence and waves, can be simulated and the flow velocity can be adjusted
from 0.2 m/s to 1.8 m/s.

Figure 4.20 shows stator current signals of the electrical generator in a TST in
different health states. Figure 4.20(a) is the current signal in the healthy state, and
Figure 4.20(b) is in the imbalance fault state.

In the time domain, with the current amplitude changing continuously due to the
uncertainties of the marine environment, it is difficult to distinguish which one is in the
fault state. Thus, for this type of condition monitoring to be effective, small physical
changes in the machine must be detected from a very noisy signal. Detection is easier
if a signal feature amplifies the change and provides a reliable signature. The presence
or absence of phase coupling between the signal's frequency components can provide
such a signature. Since the power spectral density (PSD) discards phase information,

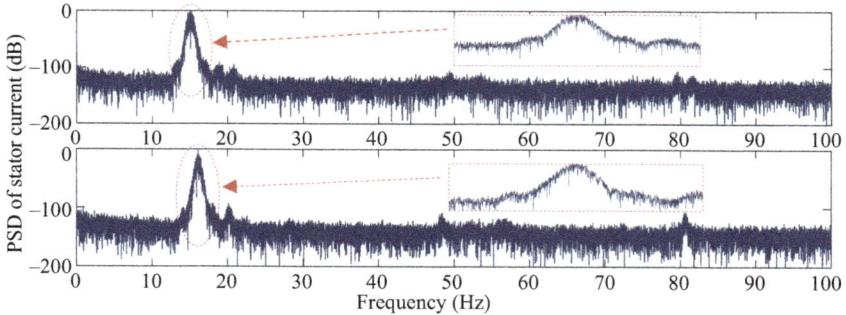

Figure 4.21   Stator current power spectrum density (PSD): healthy condition (upper PSD) and imbalanced rotor blades (lower PSD)

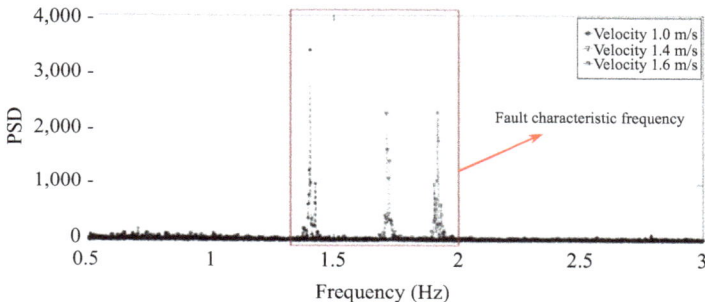

Figure 4.22   The spectrum of the fault characteristic signals under different water flow velocities

it cannot detect the presence of phase coupling; as illustrated in Figures 4.21 and 4.22, various literature have proposed different approaches to overcome this issue.

In [61], a wavelet threshold denoising-based imbalance fault detection method for MCTs was proposed to reduce the interference and detect the imbalance fault under different flow velocity conditions. The proposed approach contains three parts:

- Step 1: The wavelet threshold denoising is applied to deal with the interferences caused by turbulence and waves under different flow velocity conditions.
- Step 2: The stator current signal in the time domain is transformed to the time–frequency domain by the Hilbert transform.
- Step 3: The PCA is used, and the imbalance fault is detected by the computation of statistics indices in the principal and residual subspaces.

This approach was assessed through experimental validation. The results under different flow velocity conditions with Q statistics have shown satisfactory imbalance fault detection with false alarm and false negative rates less than 1% and 5%, respectively. Still, in the context of blade imbalance detection, in [85], the authors propose to exploit an approach based on a high-order statistics analysis (HOS). In

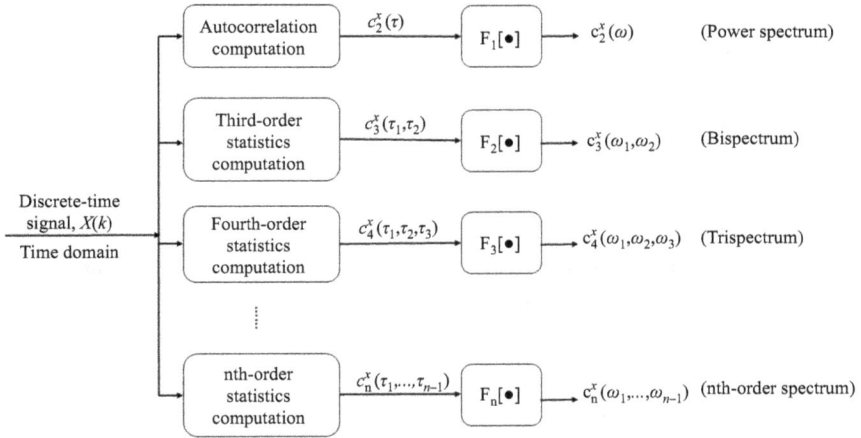

Figure 4.23   *The higher-order spectra classification map of a discrete signal X(k) [86]*

the HOS-based approach, depicted in Figure 4.23, the bispectrum is investigated. The bispectrum is defined as the frequency domain representation of the third-order cumulant or moment, while the bicoherence is the normalized representation of the bispectrum, and these two representations are indexed as processes that deviate from Gaussianity and linearity can conveniently be studied by using higher-order statistics, particularly the bispectrum and bicoherence. The bispectrum performs well detecting and quantifying the quadratic phase coupling (QPC) [86–88]. Meanwhile, it describes statistical links between signal frequency components. It is defined as:

$$B(f_1, f_2) \triangleq DDFT[c_3(\tau_1, \tau_2)] \equiv E\{X(f_1)X(f_2)X^*(f_1 + f_2)\} \tag{4.11}$$

where DDFT is the double-discrete Fourier transformation. The expression of the bispectrum shows that it is a complex quantity having magnitude and phase, and it can be plotted as a function of two independent frequency variables, f_1 and f_2. Since the bispectrum detects phase coupling, its magnitude varies with the power in the signal and therefore is not convenient for detection purposes, so the bicoherence, which is the normalized bispectrum, overcomes this problem. Under practical conditions, when handling real stator current data from monitored electromechanical systems, the signal mean is first computed and deduced from the signal. Following the mathematical bases of HOS analysis, diverse order correlation functions may be calculated for the random process as listed below:

$$\mu_x = E\{x(t)\} = 0 \ (or, \ a \ constant) \tag{4.12}$$

$$R_{xx}(\tau) = E\{x^*(t)x(t + \tau)\} \tag{4.13}$$

$$R_{xxx}(\tau_1, \tau_2) = E\{x^*(t)x(t + \tau_1)x(t + \tau_2)\} \tag{4.14}$$

$$R_{xxx\ldots}(\tau_1, \tau_2, \ldots, \tau_n) = E\{x^*(t)x(t + \tau_1)x(t + \tau_2)\ldots x(t + \tau_n)\} \tag{4.15}$$

The current $i_s(t)$ expressed by (4.9) can be represented by its analytical signal, denoted $\hat{i}_s(t)$ as:

$$\hat{i}_s(t) = A_t e^{j(\psi t + \varphi)}, \tag{4.16}$$

and the DFT of $i_s(t)$ is given by:

$$I_s(\omega) = \frac{A_t}{2}\delta(\omega - \psi)e^{j\varphi} + \frac{A_t}{2}\delta(\omega + \psi)e^{-j\varphi}, \tag{4.17}$$

where $\omega = 2\pi f$, and $\delta(\cdot)$ is the Kronecker delta function. By ignoring negative frequencies, $I_s(\omega)$ can be expressed as:

$$I_s(\omega) = \frac{A_t}{2}\delta(\omega - \psi)e^{j\varphi}. \tag{4.18}$$

Therefore, each frequency component $I_s(\omega_1)$, $I_s(\omega_1, \omega_2)$ produces one straight line in the (ω_1, ω_2) plane. For instance, $I_s(\omega_1)$ will be represented by one line at $\omega_1 = \psi$. Then, the bispectrum is equal to (4.19):

$$\begin{aligned}
\hat{B}(\omega_1, \omega_2) &= E\left\{I_s(\omega_1)I_s(\omega_2)I_s^*(\omega_1 + \omega_2)\right\} \\
&\approx \frac{1}{8}\left(A_t\delta\left(\omega_1 - \psi\right)e^{j\varphi}\right) \\
&\times \left(A_t\delta\left(\omega_2 - \psi\right)e^{j\varphi}\right) \\
&\times \left(A_t\delta\left(\omega_1 + \omega_2 - \psi\right)e^{-j\varphi}\right)
\end{aligned} \tag{4.19}$$

and the bispectrum of the current $i_s(t)$ is depicted in Figure 4.24, where the nonzero values are indexed in the (ω_1, ω_2) plane by points given in (4.20):

$$\begin{cases}
\omega_1 = \psi \\
\omega_2 = \psi \\
\omega_1 + \omega_2 = \psi
\end{cases} \tag{4.20}$$

The bispectrum $\hat{B}(\omega_s, \omega_s)$ or $\hat{B}(f_s, f_s)$ is used as a condition indicator to distinguish between the faulty and the healthy blade states. The bispectrum magnitude response, $\hat{B}(f_s, f_s)$, shows, in this case, its clear ability to detect the blade imbalance. Further

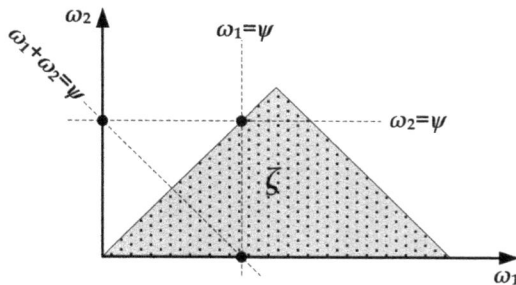

Figure 4.24   Theoretical bispectrum of tidal stream turbine stator current signal

information can be retrieved from the same stator current data by expanding the analysis to the QPC behavior.

The bicoherence or the normalized bispectrum in (4.21) is a measure of the degree of QPC that appears in a signal or between two signals' frequency components. As above-mentioned, PC is the estimate of the amount of energy in every potential pair of frequency components, $f1, f2$, which fulfills the definition of the QPC phase of the component at f_3, which is $f_1 + f_2$, equals phase of f_1 + phase of f_2 [89,90]:

$$bic(f_1, f_2) = \frac{|B(f_1, f_2)|^2}{X(f_1) X(f_2) X(f_1 + f_2)} \qquad (4.21)$$

Figures 4.25(a), 4.26(a), 4.27(a), and 4.25(b), 4.26(b), and 4.27(b) depict the stator current bispectrum amplitude, the normalized bispectrum, and the diagonal slice of the normalized bispectrum for healthy and imbalance case, respectively. The healthy case has the least quadratic nonlinear interactions (NITs) frequency among the other cases. The bifrequency points indexed by $(f_s, 0)$, (f_s, f_s) represent the highest bispectral peaks. The rotor imbalance case revealed an increased frequency interaction along f_s, this can be seen at the bifrequency coordinates $(f_s, 0)$, (f_s, f_s). An interesting observation is the high bispectral peak at (f_s, f_s) when compared with the healthy case. One can notice from the diagonal slice of the bispectrum (DSB) for the faulty case, depicted in Figure 4.27(b), that the peak amplitude is more than two times higher compared to the healthy case. A deep insight into (4.19) shows the presence of peaks in the bispectrum. Clearly, it can be identified by the nonzero products among the three terms of this equation [85]. The experimental results show that it can be seen the existence of two peaks located at (f_s, f_s) and $(f_s, 0)$, this corresponds to (15.63, 15.63) Hz and (15.63, 0) Hz.

4.4.4   Image processing-based imbalance fault detection

For image processing in undersea and TST fault and detection and diagnosis context, one of the major drawbacks is the restricted amount of available light. In the meantime, the tide can generate a swift stream of up to several meters per second, forcing the TST to rotate rapidly. Due to the aforementioned two factors, photographs captured underwater are fuzzy and dim. For instance, Figure 4.28(a) depicts an example of an image captured from the undersea environment. In the worst-case scenario, these images render human recognition impossible. Moreover, static and rotating TST pictures are shown in Figure 4.28(b). It shows clearly that TST photos are greatly affected by motion blur.

In [91], an identification method of blade attachment based on sparse autoencoder and softmax regression is suggested; the proposed approach is illustrated in Figure 4.29, this method has four steps. During the training step, some of the data is split into two groups: data without labels and data with labels.

- Step 1: The convolution kernels are pre-trained using a sparse autoencoder, which takes preprocessed unlabeled images as an input.
- Step 2: The convolved features are calculated using a convolution between the trained convolution kernels and the labeled images.

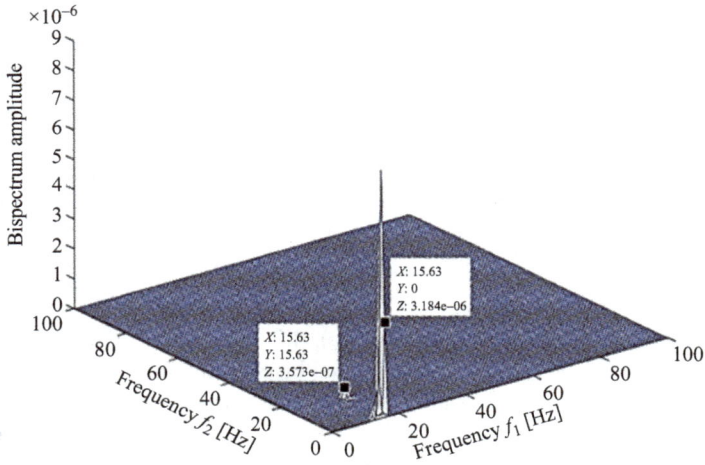

(a)

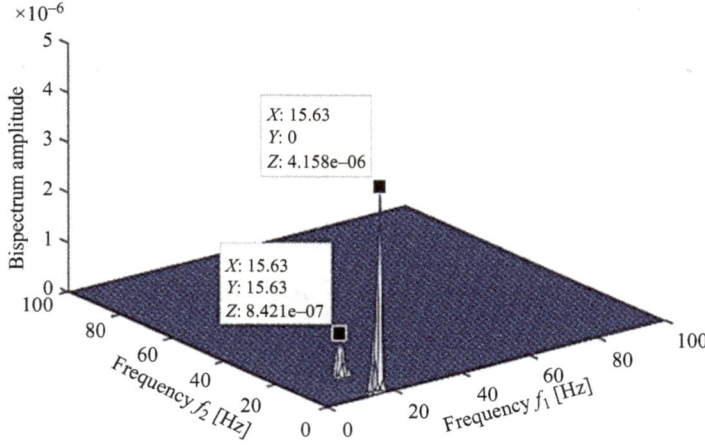

(b)

Figure 4.25   Stator current bispectrum. (a) Healthy rotor blades. (b) Imbalanced rotor blades.

- Step 3: The pooled features are calculated using the average pooling operation, which takes convolutional features as an input.
- Step 4: Uses the softmax classifier to diagnose the defects category; for this purpose, the softmax classifier uses pooled characteristics as an input.

For the training of the sparse-encoder, unlabeled images are used to extract patches. For this purpose, 500 patches of 20×20 from each unlabeled image, and arranged in matrix $X_{unlabled} = [x^1_{unlabled}, ..., x^k_{unlabled}, ...]$, where $x_{unlabel}{}^k$ is the kth column of

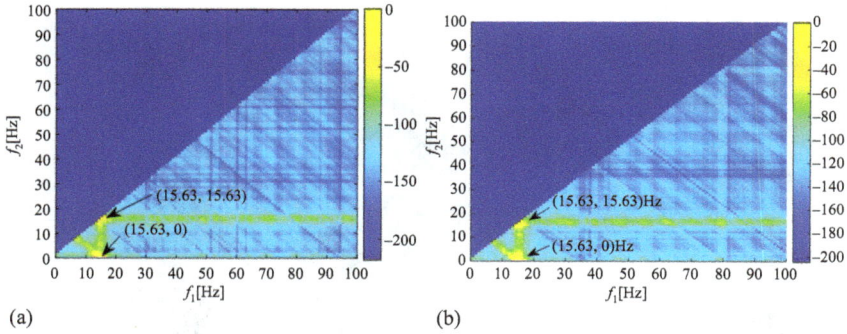

Figure 4.26   Normalized bispectrum amplitude image with scaled colors in dB. (a) Healthy rotor blades. (b) Imbalanced rotor blades.

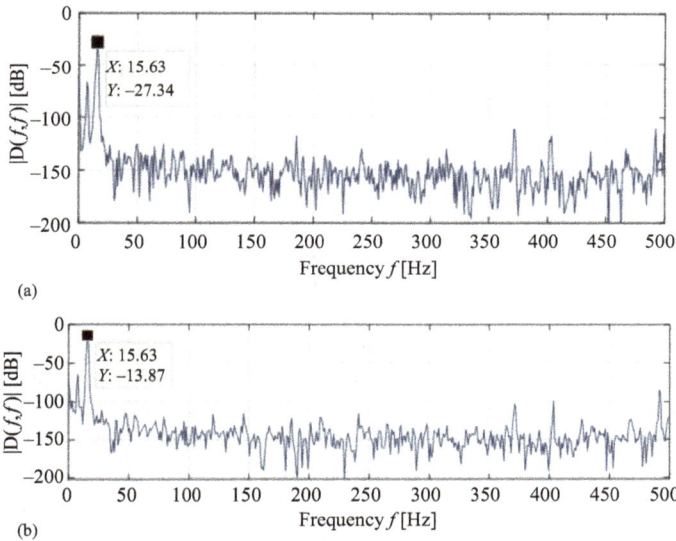

Figure 4.27   Diagonal slice of the normalized bispectrum amplitude. (a) Healthy rotor blades. (b) Imbalanced rotor blades.

$X_u nlable$. The $X^*_{unlabel}$ is obtained by removing the mean of $X_{unlabel}$. Then the covariance matrix C_X, the eigenvectors of the covariance matrix of $X^*_{unlabel}$, and the diagonal matrix of eigenvalues are calculated using as follows:

$$x^{*k}_{unlabel} = x^k_{unlabel} - \frac{1}{m}\sum_{i=1}^{m} x^i_{unlabel} \tag{4.22}$$

$$C_X = \frac{1}{m}X^*_{unlabel}\left(X^*_{unlabel}\right)^{\mathrm{T}}, \tag{4.23}$$

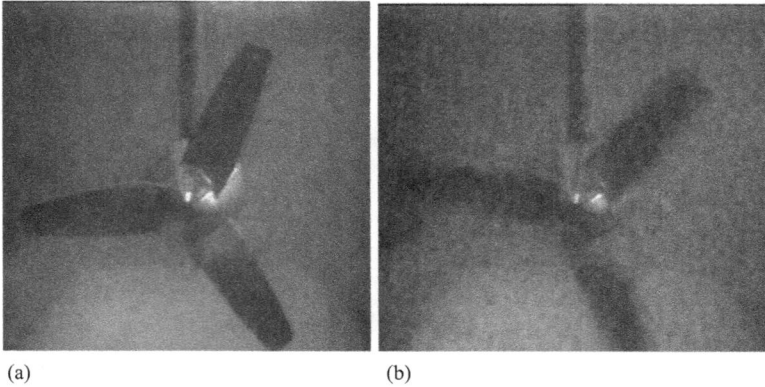

(a)   (b)

Figure 4.28   Static and rotating TST pictures. (a) Static TST. (b) Rotating TST.

and finally, the zero-phase component (ZCA) whitening [92] is used to calculate matrix $X_{whitening}$, which is given by:

$$X_{whitening} = U(S + I)^{-\frac{1}{2}} X^{*}_{unlabel},$$ (4.24)

where m is the number of samples, S is the diagonal matrix of eigenvalues, U is the eigenvectors of C_X, and ε is the regularization parameter.

For the training phase, the convolutional kernels and parameters softmax classifier are simultaneously trained, as illustrated in Figure 4.29. The sparse-autoencoder (SA) neural network is depicted in Figure 4.30, it is used for feature extraction and dimensionality reduction and has been used in various applications such as image recognition and natural language processing. The spares-autoencoder has three layers: the input layer, hidden layer, and output layer which are as denoted (L_1), (L_2), and (L_3), respectively, and the "+1" is the threshold. The sparse-autoencoder applies the backpropagation by setting the target values (the outputs) to be equal to the inputs, this means that the sparse-autoencoder can learn features by encoding and decoding training data. The input of the SA neural network is the matrix $X_{whitening}$, where $x^k \in \mathbb{R}^n$, with $X_{whitening} = [x^1, \ldots, x^k]$ and n is the number of pixels of each patch. The input $Z^{(2)}$ of action function of the hidden layer is obtained by making matrix multiplication between input matrix $X_{whitening}$ and $W^{(1)}$. The output $A^{(2)}$ is the result of the action function of the hidden layer. The detailed formulation is given by the following equations:

$$z_j^{(2)} = \sum_{i=1}^{S_1} W_{ji}^{(1)} x_i + b_j^{(1)}$$ (4.25)

$$a_j^{(2)} = f_1\left(z_j^{(2)}\right) = \frac{1}{1 + \exp\left(-z_j^{(2)}\right)}$$ (4.26)

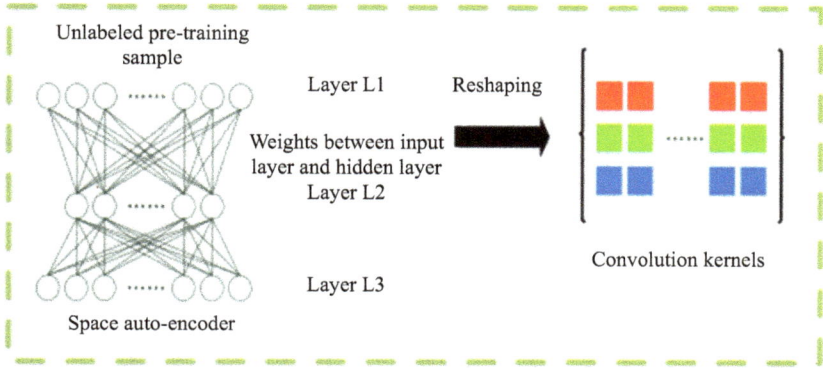

Step 2: Pre-training convolution kernels

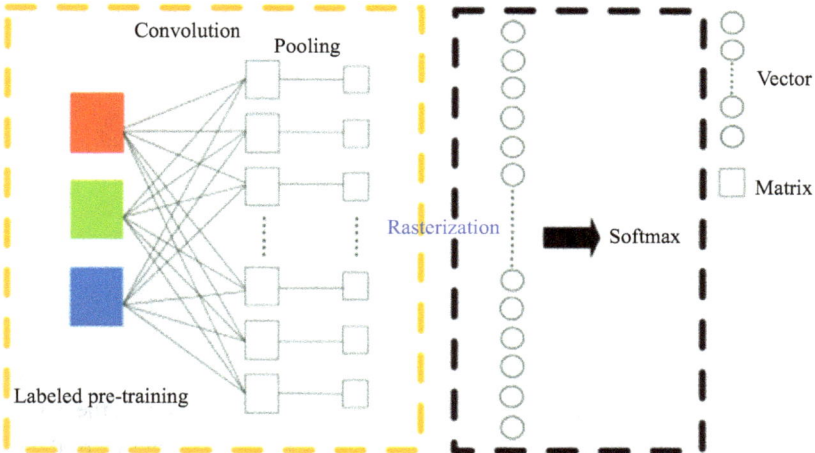

Step 3: Feature extraction through convolutional kernel trained by SA, and pooling operation

Step 4: Classification by Softmax.

Figure 4.29 *Frame of the method based on sparse autoencoder and softmax regression*

$$z_i^{(3)} = \sum_{j=1}^{S_2} W_{ij}^{(2)} a_j^{(2)} + b_i^{(2)} \tag{4.27}$$

$$a_i^{(3)} = f_2\left(z_i^{(3)}\right) = t z_i^{(3)} \tag{4.28}$$

For $i = 1, \ldots, s_1$ and $j = 1, \ldots, s_2$, $W_{ji}^{(1)}$ denotes the weight connecting the ith element of L_1 and the jth element of L_2, with the sigmoid function as an activation function, and $W_{ij}^{(2)}$ denotes the weight connecting the jth element of L_2 and the ith element of

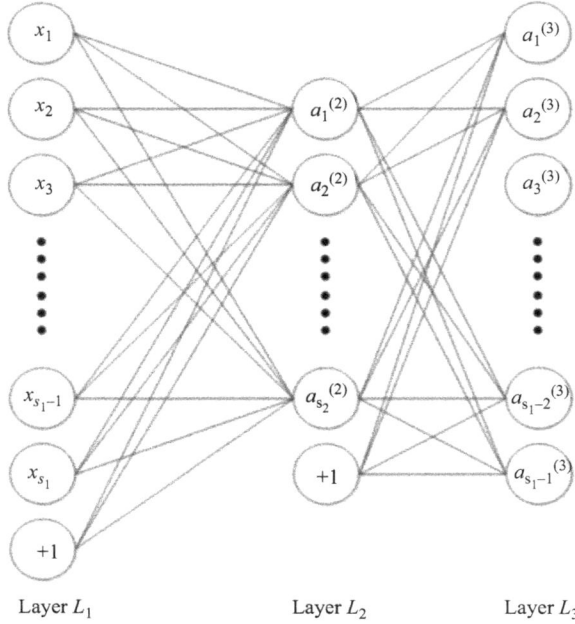

Layer L_1 Layer L_2 Layer L_3

Figure 4.30   Sparse-autoencoder neural network structure

L_3, with the proportional function as an activation function. The x_i is the ith element of vector x, $z_j^{(2)}$ and $a_j^{(2)}$ are the output of the linear transformation and the activation function in the jth elements of L_2, respectively, $z_i^{(3)}$ and $a_i^{(3)}$ are the output of the linear transformation and the activation function in the ith element of L_3, respectively, and t is the proportionality coefficient. For the feature extraction, the trained convolution kernels produce the convolutional feature maps, and the convolutional feature maps are pooled to retrieve the pooled features. Convolution of each labeled image with the pre-trained kernels yields to a different feature value at each position in the labeled image. Figure 4.31 depicts the convolution procedure. The convolutional kernel multiplies a pixel with any adjacent pixels, and the product is then added to achieve the convolved feature. This operation is repeated several times up until every pixel of the image has been covered, this operation will be done several times. After the convolution operation, it is necessary to extract the features from the outputs; for this purpose, the pooling procedure is used to minimize the size of convolutional feature maps by removing features from the output of convolution, and the pooling operation is illustrated in Figure 4.32. This will likewise adhere to the identical procedure of sliding over the convoluted features with a given pool size. There are two available forms of pooling: maximum pooling and average pooling. Due to the simplicity of the features, a standard pooling procedure was investigated. After collecting the

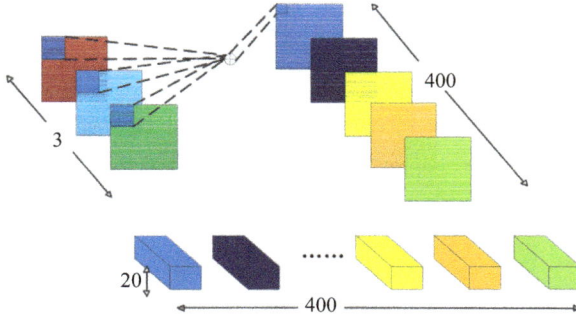

Figure 4.31 Convolution operation

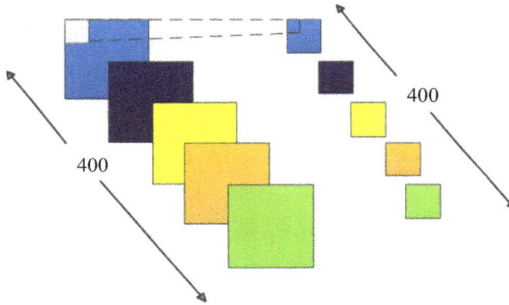

Figure 4.32 Pooling operation

pooling features, they are passed to the softmax classifier and processed according to (4.29):

$$
h_\theta\left(x^{(i)}\right) = \begin{bmatrix} p(y^{(i)} = 1|x^{(i)};\theta) \\ p(y^{(i)} = 2|x^{(i)};\theta) \\ \vdots \\ p(y^{(i)} = k|x^{(i)};\theta) \end{bmatrix} = \frac{1}{\sum_{j=1}^{k} e^{\theta_j x^{(i)}}} \begin{bmatrix} e^{\theta_1 x^{(i)}} \\ e^{\theta_2 x^{(i)}} \\ \vdots \\ e^{\theta_k x^{(i)}} \end{bmatrix},
\qquad (4.29)
$$

where θ is a parameter of softmax, x is the input of softmax, $y^{(I)}$ is the label of $x^{(I)}$, and $p(y^{(i)} = 1|x^{(i)};\theta)$ is the probability that x is classified into category 1.

This approach was assessed using the same experimental test bench, with four different configurations as depicted in Figure 4.33. Sampled images consist of unlabeled pre-training, labeled training, and testing images. The detailed information is shown in Tables 4.5, 4.6, and 4.7. This analysis simultaneously compares the training techniques of CNN and SA-softmax. According to Table 4.8, the findings demonstrate that the identification approach based on sparse autoencoder and softmax regression accurately diagnoses the various attachment degrees.

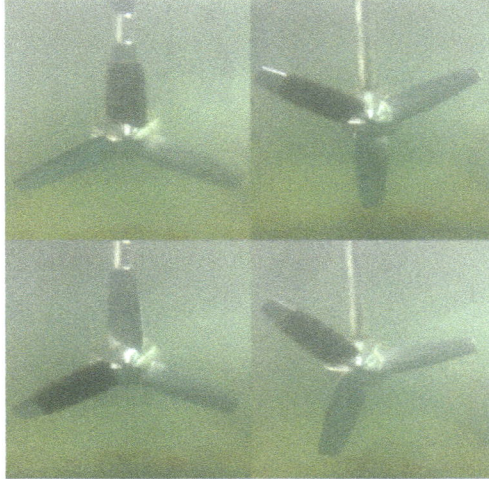

Figure 4.33   Single blade with different degrees attachment

Table 4.5   Diagnostic category label

Attachment degree	Classifier labels
0%, 0%, 0–5%	1
0%, 0%, 5–20%	2
0%, 0%, 20–40%	3
0%, 0%, 40–60%	4
0%, 0%, 60–90%	5
0%, 60–90%, 60–90%	6
60–90%, 60–90%, 60–90%	7

Table 4.6   Detail of dataset

Dataset	Number
Unlabeled pre-training image	160
Labeled training image	420
Testing image	280

Table 4.7   The parameters for the method

Parameters	Value
ε	0.1
m	80,000
λ_1	0.003
β	3
ρ	0.1
λ_2	0.0001
t	1
Hidden size	400

Table 4.8   Diagnostic results based on different methods

Diagnosis method	Average accuracy
SA+softmax	98.214%
CNN	97.500%

4.5   Summary and conclusion

Tidal turbines have undergone massive technological developments over the past decades. From simple turbines equipped with a generator to complex power plants driving the global sustainability plan, it is well known that the demand for condition monitoring processes increases as technology advances. Fault detection and diagnosis of tidal turbines is a challenging task of paramount importance due to the systems' offshore location, the submerged or immersed conditions, and the harsh operational environment. In this regard, there is a clear need for high reliability, given the severe limitations of maintenance access. This chapter has discussed the imbalance fault in the context of tidal turbines and has highlighted some methods for examining imbalance fault detection based on emerging techniques based on advanced signal processing and image processing.

References

[1]   Amy P and Rémi G. *Ocean Energy Key Trends and Statistics 2020*; 2021.

[2]   Nash S and Phoenix A. A review of the current understanding of the hydro-environmental impacts of energy removal by tidal turbines. *Renewable and Sustainable Energy Reviews*. 2017;80:648–662. https://www.sciencedirect.com/science/article/pii/S1364032117309346.

[3]   Flambard J, Amirat Y, Feld G, *et al.* River and estuary current power overview. *Journal of Marine Science and Engineering.* 2019;7(10):365. https://www.mdpi.com/2077-1312/7/10/365.

[4]   Zhou Z, Benbouzid M, Charpentier JF, *et al.* Developments in large marine current turbine technologies – a review. *Renewable and Sustainable Energy Reviews.* 2017;71:852–858. https://www.sciencedirect.com/science/article/pii/S1364032116311698.

[5]   Touimi K, Benbouzid M, and Tavner P. Tidal stream turbines: with or without a gearbox? *Ocean Engineering.* 2018;170:74–88. https://www.sciencedirect.com/science/article/pii/S0029801818313155.

[6]   Faris E, David M, Joao Amaral T, *et al.* Establishment of condition based maintenance for tidal turbines. In: *2014 27th International Congress of Condition Monitoring and Diagnostic Engineering (COMADEM 2014)*; 2014. p. 286–291.

[7]   Mekri F, Elghali SB, and Benbouzid MEH. Fault-tolerant control performance comparison of three- and five-phase PMSG for marine current turbine applications. *IEEE Transactions on Sustainable Energy.* 2013;4(2):425–433.

[8]   Elasha F, Mba D, and Teixeira JA. Condition monitoring philosophy for tidal turbines. *International Journal of Performability Engineering.* 2014;10(5): 521–534.

[9]   Elasha F, Mba D, and Texira JA. Condition monitoring philosophy for tidal turbines. *International Journal of Performability Engineering.* 2014;10(5): 521–534.

[10]   Consortium R. *Advanced Monitoring, Simulation and Control of Tidal Devices in Unsteady, Highly Turbulent Realistic Tide Environments*; 2020.

[11]   Gašperin M, Juričić, Boškoski P, *et al.* Model-based prognostics of gear health using stochastic dynamical models. *Mechanical Systems and Signal Processing.* 2011;25(2):537–548. https://www.sciencedirect.com/science/article/pii/S0888327010002396.

[12]   Richard A and Dara O. Choosing the best vibration sensor for wind turbine condition monitoring. *Analog Device Technical Article.* 2020;54(3):1–6.

[13]   Wang W and Jianu OA. A smart sensing unit for vibration measurement and monitoring. *IEEE/ASME Transactions on Mechatronics.* 2010;15(1):70–78.

[14]   Tavner P, Ran L, Penman J, *et al. Condition Monitoring of Rotating Electrical Machines.* 2nd ed. IET, UK; 2008.

[15]   Schoen RR, Habetler TG, Kamran F, *et al.* Motor bearing damage detection using stator current monitoring. *IEEE Transactions on Industry Applications.* 1995;31(6):1274–1279.

[16]   Frosini L, Harlişca C, and Szabó L. Stator current and motor efficiency as indicators for different types of bearing fault in induction motor. *IEEE Transactions on Industrial Electronics.* 2010;57(1):244–251.

[17]   Ibrahim A, Badaoui ME, Guillet F, *et al.* A new bearing fault detection method in induction machines based on instantaneous power factor. *IEEE Transactions on Industrial Electronics.* 2008;55(12):4252–4259.

[18]   Tavner PJ, Xiang J, and Spinato F. Reliability analysis for wind turbines. *Journal of Wind Energy.* 2006;10(1):1–18.

[19]   Nandi S, Toliyat HA, and Xiaodong L. Condition monitoring and fault diagnosis of electrical motors – a review. *IEEE Transactions on Energy Conversion*. 2005;20(4):719–729.

[20]   Benbouzid MEH. A review of induction motors signature analysis as a medium for faults detection. *IEEE Transactions on Industrial Electronics*. 2000;47(5):984–993.

[21]   Elbouchikhi E, Choqueuse V, and Benbouzid MEH. Induction machine faults detection using stator current parametric spectral estimation. *Mechanical Systems and Signal Processing*. 2015;52–53:447–464.

[22]   Kia SH, Henao H, and Capolino GA. A high-resolution frequency estimation method for three-phase induction machine fault detection. *IEEE Transactions on Industrial Electronics*. 2007;54(4):2305–2314.

[23]   Stoica P and Moses R. *Introduction to Spectral Analysis*. Prentice-Hall, Hoboken, NJ; 1997.

[24]   Elbouchikhi E, Choqueuse V, and Benbouzid M. Induction machine bearing faults detection based on a multi-dimensional MUSIC algorithm and maximum likelihood estimation. *ISA Transactions*. 2016;63:413 – 424.

[25]   Blodt M, Bonacci D, Regnier J, *et al*. On-line monitoring of mechanical faults in variable-speed induction motor drives using the Wigner distribution. *IEEE Transactions on Industry Applications*. 2008;55(2):522–533.

[26]   Blodt M, Regnier J, and Faucher J. Distinguishing load torque oscillations and eccentricity faults in induction motors using stator current Wigner distributions. *IEEE Transactions on Industry Applications*. 2009;45(6):1991–2000.

[27]   Antonino-Daviu JA, Riera-Guasp M, Pineda-Sanchez M, *et al*. A critical comparison between DWT and Hilbert–Huang-based methods for the diagnosis of rotor bar failures in induction machines. *IEEE Transactions on Industry Applications*. 2009;45(5):1794–1803.

[28]   Cusido JC, Romeral L, Ortega JA, *et al*. Fault detection in induction machines using power spectral density in wavelet decomposition. *IEEE Transactions on Industrial Electronics*. 2008;55(2):633–643.

[29]   Riera-Guasp M, Antonio-Daviu JA, Roger-Folch J, and Molina Palomares MP. The use of the wavelet approximation signal as a tool for the diagnosis of rotor bar failure. *IEEE Transactions on Industry Applications*. 2008;44(3):716–726.

[30]   Kia SH, Henao H, and Capolino GA. Diagnosis of broken-bar fault in induction machines using discrete wavelet transform without slip estimation. *IEEE Transactions on Industry Applications*. 2009;45(4):1395–1404.

[31]   Mandic DP, u Rehman N, Wu Z, *et al*. Empirical mode decomposition-based time-frequency analysis of multivariate signals: the power of adaptive data analysis. *IEEE Signal Processing Magazine*. 2013;30(6):74–86.

[32]   Tavner PJ. Review of condition monitoring of rotating electrical machines. *IET Electric Power Applications*. 2008;2(4):215–247.

[33]   Stack JR, Harley RG, and Habetler TG. An amplitude modulation detector for fault diagnosis in rolling element bearings. *IEEE Transactions on Industrial Electronics*. 2004;51(5):1097–1102.

[34]   Riley CM, Lin BK, Habetler TG, *et al.* A method for sensorless on-line vibration monitoring of induction machines. *IEEE Transactions on Industry Applications.* 1998;34(6):1240–1245.

[35]   Popa LM, Jensen BB, Ritchie E, *et al.* Condition monitoring of wind generators. In: *38th IAS Annual Meeting on Conference Record of the Industry Applications Conference,* 2003, vol. 3; 2003. p. 1839–1846.

[36]   Benbouzid M, editor. *Signal Processing for Fault Detection and Diagnosis in Electric Machines and Systems.* Institution of Engineering and Technology; 2020.

[37]   Khamoudj CE, Tayeb FBS, Benatchba K, *et al.* A learning variable neighborhood search approach for induction machines bearing failures detection and diagnosis. *Energies.* 2020;13(11):2953.

[38]   Tang G and Tian T. Compound fault diagnosis of rolling bearing based on singular negentropy difference spectrum and integrated fast spectral correlation. *Entropy.* 2020;22(3):367.

[39]   Mao W, Sun B, and Wang L. A new deep dual temporal domain adaptation method for online detection of bearings early fault. *Entropy.* 2021;23(2):162.

[40]   Peimankar A and Puthusserypady S. DENS-ECG: a deep learning approach for ECG signal delineation. *Expert Systems with Applications.* 2021;165:113911.

[41]   Zhang Y, Li X, Gao L, *et al.* A new subset based deep feature learning method for intelligent fault diagnosis of bearing. *Expert Systems with Applications.* 2018;110:125–142.

[42]   Berghout T, Mouss LH, Bentrcia T, *et al.* A deep supervised learning approach for condition-based maintenance of naval propulsion systems. *Ocean Engineering.* 2021;221:108525.

[43]   Lu S. *IET Science, Measurement & Technology.* 2021;15:551–561(10). https://digital-library.theiet.org/content/journals/10.1049/smt2.12056.

[44]   Li H, Huang J, Yang X, *et al.* Fault diagnosis for rotating machinery using multiscale permutation entropy and convolutional neural networks. *Entropy.* 2020;22(8):851.

[45]   Xu Z, Li C, and Yang Y. Fault diagnosis of rolling bearing of wind turbines based on the variational mode decomposition and deep convolutional neural networks. *Applied Soft Computing.* 2020;95:106515.

[46]   Qiao M, Yan S, Tang X, *et al.* Deep convolutional and LSTM recurrent neural networks for rolling bearing fault diagnosis under strong noises and variable loads. *IEEE Access.* 2020;8:66257–66269.

[47]   Habbouche H, Amirat Y, Benkedjouh T, *et al.* Bearing fault event-triggered diagnosis using a variational mode decomposition-based machine learning approach. *IEEE Transactions on Energy Conversion.* 2022;37(1):466–474.

[48]   Cao W, Yan J, Jin Z, *et al.* Image denoising and feature extraction of wear debris for online monitoring of planetary gearboxes. *IEEE Access.* 2021;9: 168937–168952.

[49]   Jaffery ZA. *Thermal Image Based Monitoring of PV Modules and Solar Inverters.* IET Press; 2021.

[50]   Kang HS, Lee JY, Choi S, *et al.* Smart manufacturing: past research, present findings, and future directions. *International Journal of Precision*

Engineering and Manufacturing-Green Technology. 2016;3(1):111–128. https://doi.org/10.1007/s40684-016-0015-5.

[51] Herrmann C, Schmidt C, Kurle D, *et al.* Sustainability in manufacturing and factories of the future. *International Journal of Precision Engineering and Manufacturing-Green Technology.* 2014;1(4):283–292. https://doi.org/10. 1007/s40684-014-0034-z.

[52] Rusk N. Deep learning. *Nature Methods.* 2016 Jan;13(1):35–35. https://doi. org/10.1038/nmeth.3707.

[53] LeCun Y, Bengio Y, and Hinton G. Deep learning. *Nature.* 2015;521(7553): 436–444. https://doi.org/10.1038/nature14539.

[54] Hochreiter S and Schmidhuber J. Long short-term memory. *Neural Computation.* 1997;9(8):1735–1780. https://doi.org/10.1162/neco.1997.9.8.1735.

[55] Bai J and Feng XC. Fractional-order anisotropic diffusion for image denoising. *IEEE Transactions on Image Processing.* 2007;16(10):2492–2502.

[56] Fukushima N, Sugimoto K, and Kamata SI. Guided image filtering with arbitrary window function. In: *2018 IEEE International Conference on Acoustics, Speech and Signal Processing (ICASSP)*; 2018. p. 1523–1527.

[57] Torres-Huitzil C. Fast hardware architecture for grey-level image morphology with flat structuring elements. *IET Image Processing.* 2014;8: 112–121(9). https://digital-library.theiet.org/content/journals/10.1049/iet-ipr. 2013.0224.

[58] Ren Z, Fang F, Yan N, *et al.* State of the art in defect detection based on machine vision. *International Journal of Precision Engineering and Manufacturing-Green Technology.* 2022;9(2):661–691. https://doi.org/10.1007/s40684-021-00343-6.

[59] Gonzalez RC and Woods RE. *Digital Image Processing.* Prentice-Hall, Hoboken, NJ; 2007.

[60] Flambard J, Amirat Y, Feld G, *et al.* River and estuary current power overview. *Journal of Marine Science and Engineering.* 2019;7(10). https://www.mdpi. com/2077-1312/7/10/365.

[61] Xie T, Wang T, He Q, *et al.* A review of current issues of marine current turbine blade fault detection. *Ocean Engineering.* 2020;218:108194. https://www.sciencedirect.com/science/article/pii/S0029801820311215.

[62] Titah-Benbouzid H and Benbouzid M. Marine renewable energy converters and biofouling: a review on impacts and prevention. In *EWTEC 2015*; 2015. p. Paper–09P1.

[63] Zhang D, Qian L, Mao B, *et al.* A data-driven design for fault detection of wind turbines using random forests and XGboost. *IEEE Access.* 2018;6: 21020–21031.

[64] Jiang G, Xie P, He H, *et al.* Wind turbine fault detection using a denoising autoencoder with temporal information. *IEEE/ASME Transactions on Mechatronics.* 2018;23(1):89–100.

[65] Wang T, Qi J, Xu H, *et al.* Fault diagnosis method based on FFT-RPCA-SVM for cascaded-multilevel inverter. *ISA Transactions.* 2016;60:156–163. https://www.sciencedirect.com/science/article/pii/S0019057815002943.

[66] Zhou W, Habetler TG, and Harley RG. Bearing fault detection via stator current noise cancellation and statistical control. *IEEE Transactions on Industrial Electronics*. 2008;55(12):4260–4269.

[67] Yassine A and Mohamed B. *The Signal Demodulation Techniques*. Institution of Engineering and Technology; 2020. https://digital-library.theiet.org/content/books/10.1049/pbpo153e_ch2.

[68] Wu ZH and Huang NE. Ensemble empirical mode decomposition: a noise-assisted data analysis method. *Advances in Adaptive Data Analysis*. 2009;1:1–41.

[69] Torres ME, Colominas MA, Schlotthauer G, *et al.* A complete ensemble empirical mode decomposition with adaptive noise. In: *Proceedings of 2011 IEEE International Conference on Acoustics, Speech and Signal Processing (ICASSP)*; 2011. p. 4144–4147.

[70] Huang NE and Shen SSP. *Hilbert-Huang Transform and Its Applications*. 2nd edition, EBSCO ebook academic collection. World Scientific; 2014.

[71] Yu D, Cheng J, and Yang Y. Application of EMD method and Hilbert spectrum to the fault diagnosis of roller bearings. *Mechanical Systems and Signal Processing*. 2005;19(2):259–270.

[72] Amirat Y, Choqueuse V, and Benbouzid MEH. EEMD-based wind turbine bearing failure detection using the generator stator current homopolar component. *Mechanical Systems and Signal Processing*. 2013;41(1):667–678.

[73] Malik H and Mishra S. Artificial neural network and empirical mode decomposition based imbalance fault diagnosis of wind turbine using TurbSim, FAST and Simulink. *IET Renewable Power Generation*. 2017;11(6):889–902. https://ietresearch.onlinelibrary.wiley.com/doi/abs/10.1049/iet-rpg.2015.0382.

[74] Zhang M, Wang T, Tang T, *et al.* An imbalance fault detection method based on data normalization and EMD for marine current turbines. *ISA Transactions*. 2017;68:302–312. https://www.sciencedirect.com/science/article/pii/S0019057817302999.

[75] Malik H and Mishra S. Proximal support vector machine (PSVM) based imbalance fault diagnosis of wind turbine using generator current signals. *Energy Procedia*. 2016;90:593–603. *5th International Conference on Advances in Energy Research (ICAER)*; 2015. https://www.sciencedirect.com/science/article/pii/S1876610216314382.

[76] Zhang M, Wang T, Tang T, *et al.* Imbalance fault detection of marine current turbine under condition of wave and turbulence. In: *IECON 2016-42nd Annual Conference of the IEEE Industrial Electronics Society*. IEEE; 2016. p. 6353–6358.

[77] Golestan S, Ramezani M, Guerrero JM, *et al.* dq-Frame cascaded delayed signal cancellation-based PLL: analysis, design, and comparison with moving average filter-based PLL. *IEEE Transactions on Power Electronics*. 2015;30(3):1618–1632.

[78] Amirat Y, Benbouzid MEH, Wang T, *et al.* EEMD-based notch filter for induction machine bearing faults detection. *Applied Acoustics*. 2018;133:202–209. https://www.sciencedirect.com/science/article/pii/S0003682X17308125.

[79] Fan W, Zhou Q, Li J, *et al.* A wavelet-based statistical approach for monitoring and diagnosis of compound faults with application to rolling bearings. *IEEE Transactions on Automation Science and Engineering.* 2018;15(4): 1563–1572.

[80] Geng Z, Chen J, and Han Y. Energy efficiency prediction based on PCA-FRBF model: a case study of ethylene industries. *IEEE Transactions on Systems, Man, and Cybernetics: Systems.* 2017;47(8):1763–1773.

[81] Rafferty M, Liu X, Laverty DM, *et al.* Real-time multiple event detection and classification using moving window PCA. *IEEE Transactions on Smart Grid.* 2016;7(5):2537–2548.

[82] Sheng X, Wan S, Cheng L, *et al.* Blade aerodynamic asymmetry fault analysis and diagnosis of wind turbines with doubly fed induction generator. *Journal of Mechanical Science and Technology.* 2017;31(C):5011–5020.

[83] Chen H, At-Ahmed N, Machmoum M, *et al.* Modeling and vector control of marine current energy conversion system based on doubly salient permanent magnet generator. *IEEE Transactions on Sustainable Energy.* 2016;7(1): 409–418.

[84] Zhang M, Wang T, Tang T, *et al.* Imbalance fault detection of marine current turbine under condition of wave and turbulence. In: *IECON 2016 – 42nd Annual Conference of the IEEE Industrial Electronics Society;* 2016. p. 6353–6358.

[85] Saidi L, Benbouzid M, Diallo D, *et al.* Higher-order spectra analysis-based diagnosis method of blades biofouling in a PMSG driven tidal stream turbine. *Energies.* 2020;13(11):2888. https://www.mdpi.com/1996-1073/13/11/2888.

[86] Nikias CL and Petropulu AP. *Higher-Order Spectra Analysis a Nonlinear Signal Processing Framework.* Prentice-Hall, Hoboken, NJ; 1993.

[87] Mendel JM. Tutorial on higher-order statistics (spectra) in signal processing and system theory: theoretical results and some applications. *Proceedings of the IEEE.* 1991;79(3):278–305.

[88] Hinich MJ and Wolinsky M. Normalizing bispectra. *Journal of Statistical Planning and Inference.* 2005;130(1):405–411. Herman Chernoff: Eightieth Birthday Felicitation Volume. https://www.sciencedirect.com/science/article/pii/S0378375804002745.

[89] Nichols JM, Olson CC, Michalowicz JV, *et al.* The bispectrum and bicoherence for quadratically nonlinear systems subject to non-Gaussian inputs. *IEEE Transactions on Signal Processing.* 2009;57(10):3879–3890.

[90] Wang X, Chen Y, and Ding M. Testing for statistical significance in bispectra: a surrogate data approach and application to neuroscience. *IEEE Transactions on Biomedical Engineering.* 2007;54(11):1974–1982.

[91] Zheng Y, Wang T, Xin B, *et al.* A sparse autoencoder and softmax regression based diagnosis method for the attachment on the blades of marine current turbine. *Sensors.* 2019;19(4). https://www.mdpi.com/1424-8220/19/4/826.

[92] Krsman VD and Sarić AT. Bad area detection and whitening transformation-based identification in three-phase distribution state estimation. *IET Generation, Transmission & Distribution.* 2017;11(9):2351–2361. https://ietresearch.onlinelibrary.wiley.com/doi/abs/10.1049/iet-gtd.2016.1866.

Chapter 5

Biofouling issue in tidal stream turbines

Hosna Titah-Benbouzid[1], Haroon Rashid[1] and
Mohamed Benbouzid[1]

5.1 Introduction

A clear trend towards reorientation into renewable energies has been observed since the end of the twentieth century, in response to the beginning of scarcity of oil, the negative climate and health impacts because of carbon energies, the dangerous nature of nuclear power, and the difficulty of dealing with its waste. Fossil fuels are still the primary and the most widely used energy source, problems related to their contribution to global warming and finding new supplies are prompting us to seek cleaner and more sustainable alternatives: renewable energy [1]. Renewable energies are energy sources whose natural renewal is fast enough that they can be considered inexhaustible at human time scale. They come from cyclical or constant natural phenomena induced by stars: Sun especially, for the heat and the light, but also the Moon (tides) and the Earth (geothermal energy). Wind turbines are characterized by their performance as a function of wind speed. Current wind turbines are limited to winds of less than 90 km/h. Wind farm installation at sea is one of the development paths of this sector: it minimizes the visual and neighborhood nuisances and the load factor is better thanks to a stronger and more constant wind on the ground. In 2005–2010, the desire to develop renewable energies puts a spotlight on marine energy and tidal energy. A tidal turbine is a hydraulic turbine (underwater or in water) that uses the marine or fluvial currents kinetics energy, as a wind turbine uses the wind kinetic energy. A tidal turbine allows the transformation of kinetic energy on moving water into mechanical energy, which can be then converted into electrical energy by an alternator. The generation of low-carbon electricity is of global importance as a strategy to mitigate the impacts of climate change and to ensure energy security in the coming century. Tidal stream energy, the conversion of the kinetic energy that resides in tidal currents into electricity (typically through intercepting the flow via arrays of horizontal axis turbines [2]), is favored as a renewable energy resource for several reasons such as the predictability to provide firm renewable electricity [3–5].

[1]University of Brest, CNRS, Institut de Recherche Dupuy de Lôme, France

One of the major problems encountered in tidal turbines is the control of biofouling. It is a special case of fouling (biological fouling) defined by the colonization of any surface (living or not) in an aqueous environment by living organisms. Biofouling can be made up of hundreds of different species of living beings in the same area. Organism groups that contribute to marine biofouling include seaweeds, bivalves, crustaceans, and barnacles. Marine biofouling can be divided into two groups: (1) microfouling organisms (any substrate placed in seawater will be rapidly colonized by these organisms, which sufficiently accumulates in thicknesses to obscure the marine surfaces [6]) and (2) macrofouling organisms (this grouping includes many larger animals and plants that can be located in individuals or large colonies, such as barnacles, mussels, polychaetes and various species of bryozoa and hydroids: fouling and abrasive suspended particles growth [7]).

In this chapter, we will compare the effect of biofouling on tidal turbines with the effect of icing on wind turbines. The comparison will be based on equations that could be developed to give an example of modeling to understand the biofouling effect on speed and power generated by tidal turbines. This chapter also focuses on the kinetic energy aspects of ocean energy—namely that present in ocean currents [8]. Marine currents are increasingly being recognized as a resource to be exploited for sustainable generation of electrical power [9–12]. Oceans cover more than 70% of the Earth's surface. They offer a huge energy resource that can produce large amounts of sustainable energy. It can be exploited for the production of electricity and or freshwater (Figure 5.1).

Ocean energy has many forms and is stored partly as thermic energy, partly as kinetic energy (waves and currents) and also in chemical and biological processes. The ocean's kinetic energy can be converted into a usable form by many proposed technologies. Tidal currents, for example, can be converted to electricity using conventional horizontal axis turbine technology commonly seen in wind energy. In certain

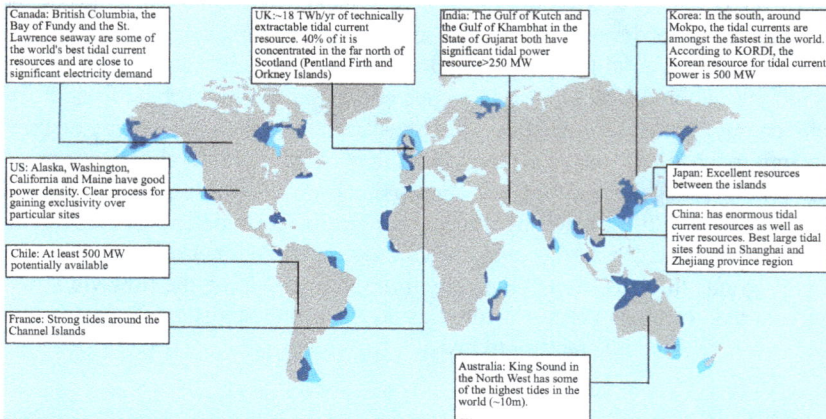

| Canada: British Columbia, the Bay of Fundy and the St. Lawrence seaway are some of the world's best tidal current resources and are close to significant electricity demand | UK:~18 TWh/yr of technically extractable tidal current resource. 40% of it is concentrated in the far north of Scotland (Pentland Firth and Orkney Islands) | India: The Gulf of Kutch and the Gulf of Khambhat in the State of Gujarat both have significant tidal power resource>250 MW | Korea: In the south, around Mokpo, the tidal currents are amongst the fastest in the world. According to KORDI, the Korean resource for tidal current power is 500 MW |

US: Alaska, Washington, California and Maine have good power density. Clear process for gaining exclusivity over particular sites

Japan: Excellent resources between the islands

China: has enormous tidal current resources as well as river resources. Best large tidal sites found in Shanghai and Zhejiang province region

Chile: At least 500 MW potentially available

France: Strong tides around the Channel Islands

Australia: King Sound in the North West has some of the highest tides in the world (~10m).

Figure 5.1   High potential areas for tidal resources [8]

regions, oceans marine currents velocities are augmented by constraining topographies resulting in appreciable velocities that could be harnessed for the production of sustainable electrical power. One such site is the Race of Alderney in the Channel Islands [13,14], which could supply 1.34 TWh/year with a large farm of turbines [8]. Marine current power is an emerging renewable technology and although technology can be transferred from wind turbines, there is a need for research specific to marine current turbines. Ng *et al.* [15] provide a comprehensive review of the horizontal axis marine current turbine research in the last ten years. Recent experimental studies on marine current turbine performance include Refs. [16–18]. Ocean energy or marine renewable energy, which to most people is an unknown renewable energy resource, has known a notable growth in the last ten years [19,20]. In a few years, a rapid expansion of Offshore Wind Farm (OWF) development is planned in part to attain the European Union (EU) target of 20% of energy generation from renewable by 2020 [21]. This will lead to OWF and Marine Renewable Energy Installations (MREIs) covering an extensive area of coastal waters, especially in the North East Atlantic and the Baltic Sea where current developments and leased areas propose to cover between 28,000 and 29,000 km^2 of sea bed [22]. The identification of favorable areas for the installation of tidal turbines is done by modeling current velocities that highlight the best deposits. The energy potential is calculated by taking into account the characteristics of deposits, the technical characteristics of machines, and the site constraints. At the beginning of April 2014, Arcouest tidal turbine was successfully raised by the DCNS (Direction of Shipbuilding, System and Service) teams and its subsidiary OpenHydro, after 4 months of real-life tests offshore at the bottom of sea in Paimpol-Brehat (Cotes d'Armor) [23]. The immersion of a first tidal turbine in 2011 continued in 2016 with the installation of a second tidal turbine at 40 m depth and 15 km off the island of Brehat but they were released from the water in April 2017 due to corrosion problems. Open Hydro continues its experiments in progress in Cherbourg (Normandy) and in the Bay of Fundy (Figure 5.2) [24].

5.2   Biofouling development

The most well-known form of biofouling is encountered in the marine environment. Biofouling colonizes ships, buoys, sonar devices, pontoons, offshore structures, oil installations, platforms, underwater cables, underwater acoustic instruments, seawater cooling systems, and marinas. Issues include increased costs, reduced speed, environmental concerns, corrosion, and safety hazards [6,25–32].

5.2.1   Characterization

Marine biofouling can be divided into two groups: (1) microfouling organisms and (2) macrofouling organisms.

- **Microfouling organisms:** These organisms are primarily bacterial and microbial in nature and rapidly colonize any substrate placed in the water. They form part of a sticky coating called biofilm. It is a film made up of bacteria, such as Thiobacilli or

Figure 5.2   Experiment in the Raz-Blanchard (Manche, Normandie) [24]

other microorganisms that, under favorable conditions, develop on a material [33]. Biofilms are a major fouling problem, accumulating in sufficient thickness to obscure surfaces and increasing the difficulties of underwater operations. They also provide a convenient food source and interface to which larger organisms, macrofouling, can adhere.

- **Macrofouling organisms:** Macrofouling organisms pose additional, more serious problems for underwater operators. These organisms (macrofoulers) represent the progression of a biofilm which can increase to a stage where it feeds a medium for the growth of algae, nacres, and other organisms [34]. This group includes many larger animals and plants that can attach individually or in large colonies, such as barnacles, mussels, polychaetes, and various species of bryozoans and hydroids. Once biological fouling growth comes in contact with renewable marine devices, it affects submerged tidal systems. Even at submerged depths of 5 m, these systems and in particular the rotor hubs (low speed parts) will be covered with fouling [28]. In the case of marine renewable energy equipment, the negative effects caused by this deposit are well known: high frictional resistance, increased weight and potential reduction in speed, and finally loss of maneuverability. This results in higher fuel consumption. A large amount of toxic waste is also generated during this process [35,36], deterioration of the coating so that corrosion, discoloration, and alteration of the electrical conductivity of the material are promoted [37], and introduction of species in environments where they were not naturally present (invasive or non-native species) [38,39].

Figure 5.3 represents the diversity and size scales of representative fouling organisms.

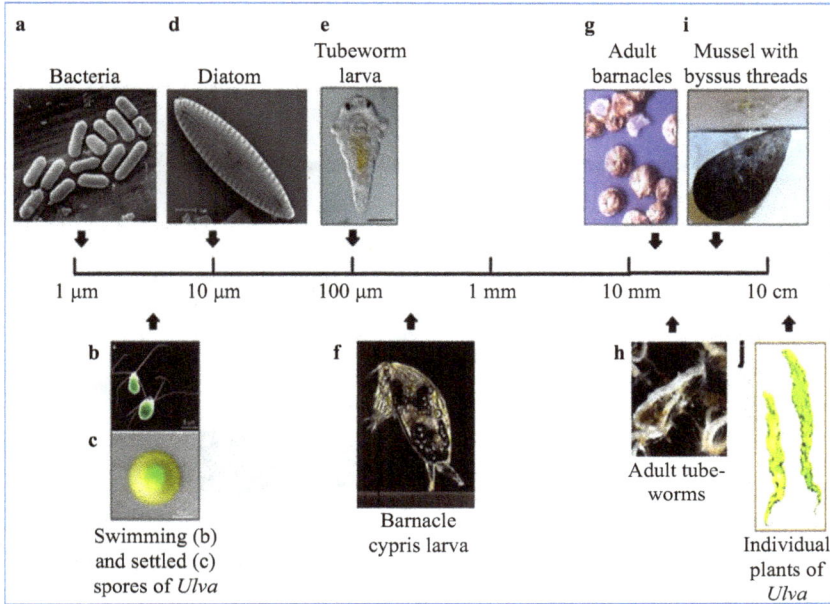

Figure 5.3   Diversity and size scales of representative fouling organisms range [34]

5.2.2   Development

In marine environment, any marine renewable energy converters (MREC) could be colonized by bodies such as bacteria, diatoms, protozoans, algae, and invertebrates [40,41]. The colonization process includes biochemical conditioning, bacterial colonization, installation of single-celled species, and the installation of multicellular species [42]. Figure 5.4 summarizes the four sequences of a typical fouling installation.

Figure 5.5 illustrates the sequential four-step process of biofouling, which is controlled by a number of physical, chemical, and biological factors. These factors are attachment, proliferation, maturation, and dispersion [43–45]. Figure 5.5 depicts the environment and particular surface on which microbial adhesion takes place, as well as the initial adherence of microorganisms to the surface, their proliferation (or else the formation of microcolonies), and the maturation of the biofilm architecture with the presence of the polymeric matrix and its dispersion [46].

The colonization process begins within days to a few weeks and biofilm covers surface. Underwater environments are ideal for biofouling as currents deliver nutrients and carry away wastes, promoting colonization by planktonic and sessile organisms. Biofouling growth rates depend on the organism, substrate, flow velocity, shear stress, and temperature [28,47,48]. Active mechanisms include electrostatic repulsion, Brownian motion, turbulent pulsations, and cell outgrowths [28,49]. Bioadhesion is the adhesion strength of biofouling on a hard surface. It depends on the

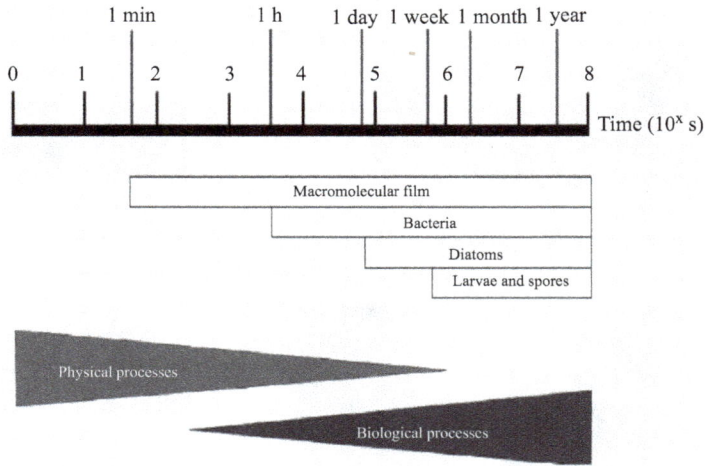

Figure 5.4   Schematized colonizing sequence leading to the establishment of a fouling community (modified from Wahl 1989) [40]

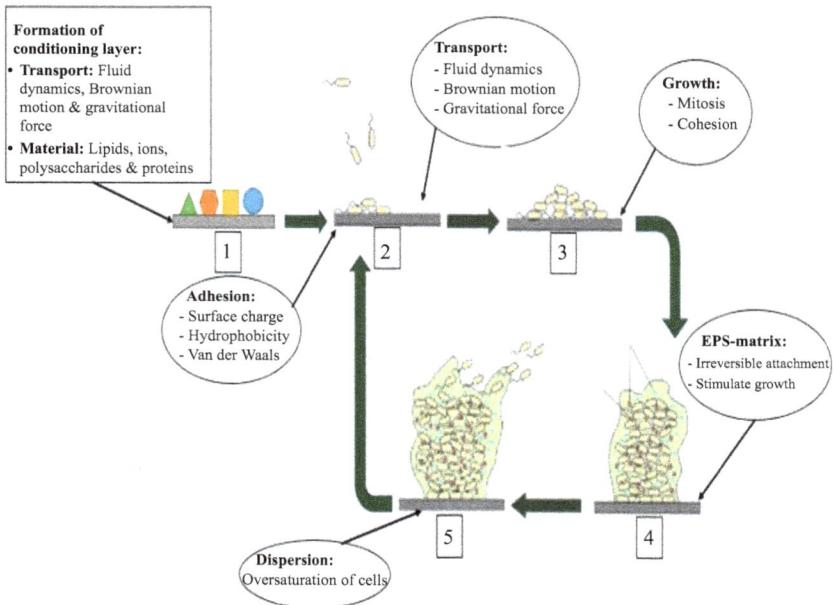

Figure 5.5   Phenomena of biofilm formation [43,46]

organism type, substrate, and separating fluid [50] owing to influences of electrostatic forces and surface wettability [51,52].

The development of a conditioning layer or film on the surface is the first phenomenon that takes place. After the fluid and surface make contact, the organic and inorganic materials present in the liquid coat the surface [53]. The base upon which a biofilm develops is the conditioning layer [46]. This layer is made up of numerous organic and inorganic particles found in the bulk fluid, such as ions, proteins, polysaccharides, and lipids. Gravitational force, fluid dynamics, and Brownian motion are all used to move these particles to the surface [54]. Physical–chemical characteristics of the surface, such as the surface charge (electrokinetics) and the surface's hydrophobicity, are significantly impacted by the formation of a conditioning layer [55,56].

The initial colonists attach to a surface through weak, reversible van der Waals bonds, which are slightly stronger than electrostatic repulsive forces. Irreversible attachment is accomplished with secretion of the extracellular polymeric substance (EPS), which exhibits a sponge-like matrix. This adhesive permanently binds the micro-organisms to one another and collectively to the surface [28,57,58]. In summary, most of the mechanisms may be and frequently are combined to form a multifactorial anti-fouling adaptation, which effectively covers the range of potential colonists [7] (Figure 5.6).

5.3   Biofouling characterization and estimation

Artificial construction equipment, found in a marine environment such as glass or fiberglass, concrete, treated wood, metal, rubber, and rigid plastic, are unfortunately affected by the proliferation of biological fouling inconvenient on materials; despite the fact that biofouling development and succession are being sought for marine fauna and flora in some reefs [7]. These biological organisms will quickly colonize the materials and can form a thickness up to several centimeters above. This phenomenon must be taken into consideration for the operation and maintenance of tidal turbines as it modifies hydrodynamics by increasing drag and therefore the resistance, which could adversely affect the performance of turbine [59,60] (Figure 5.7).

In [59,62,63], it has been reported that the attachment of barnacles to marine turbines results in a reduction in hydrodynamic performance. Hydrodynamics is significantly influenced by the density, height, thickness, and adhesion characteristics of barnacle colonized surfaces.

Demiral *et al.* [64] conducted an experimental study on the impact of artificial barnacles on ship resistance, and the results indicated that the effect of barnacle height on ship resistance is significant. For example, a 10% coverage of barnacles with a height of 5 mm caused an increase in power requirements compared to a 50% coverage of barnacles with a height of 1.5 mm.

On sea surfaces, the growth of barnacles causes roughness and increases drag. A roughness characteristic called "barnacle density" quantifies how evenly distributed the barnacles are as roughness components on a surface. Growing barnacles will

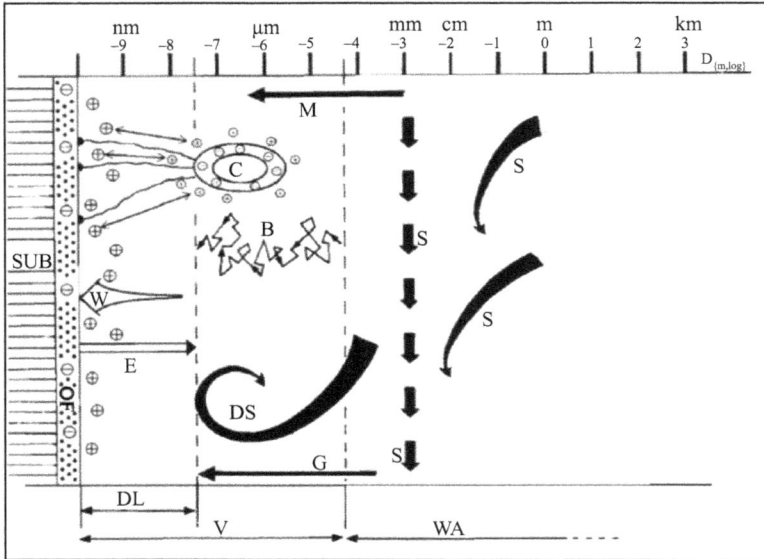

Figure 5.6 *Dominant forces as a function of the previously adsorbed macromolecular layer distance: bacterial adsorption: (B) Brownian motion; (C) bacterial cell; (DL) electrostatic double layer; (DS) downsweep (microturbulence); (E) electrostatic repulsion; (F) bacterial fibrils anchored to adsorbed macromolecules; (G) gravity (for a horizontal surface); (M) bacterial motility; (OF) organic film of adsorbed macromolecules; (S) currents and macroturbulence; (SUB) substratum; (V) viscous boundary layer; (W) Van-der-Waal's forces; (WA) water body [42]*

close the gaps between themselves, increasing the density of barnacles on a surface. When there is less room between two nearby barnacles, the drag force might rise suddenly [59,62].

The development of barnacle fouling and, consequently, barnacle density, is influenced by feeding habits. Adult barnacles have two main ways of obtaining food: either their feathery legs, known as cirri, are extended into moving water to actively or passively capture food, or they beat them rhythmically to do so [65].

The hydrodynamic forces on barnacles attached to a foil being towed by a small skiff were studied by Schultz *et al.* [66]. According to the findings, as barnacle height increases, lift force decreases and drag force increases. Additionally, increasing barnacle height has effects on sheltering and wake interaction. The loading on a single barnacle is illustrated in Figure 5.8.

An idealized model to forecast the tensile and shear stresses at the base of the barnacle is shown in (5.1) and (5.2). Depending on the adhesion strength of the

Figure 5.7   Turbine was pulled up for inspection and gave an opportunity to examine the distribution of the species and level of fouling organism's growth [61]

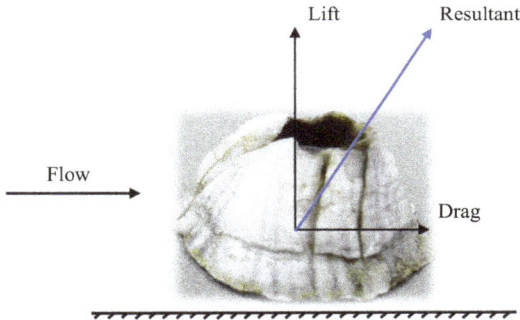

Figure 5.8   Hydrodynamic forces on a barnacle

barnacles, these stresses can be used to predict the speed at which they will separate from a surface:

$$\sigma_y = \frac{4L}{\pi d^2} + \frac{7.7D}{\pi d^2} \qquad (5.1)$$

$$\tau_{xy} = \frac{4D}{\pi d^2} \qquad (5.2)$$

where σ is the tensile stress, τ is the shear stress, D is the drag force, L is the lift force, and d is the turbine diameter.

5.3.1   Some hydrodynamic aspects to understand the effect of biofouling on tidal turbines power

In general, biofouling is a major problem that increases the structural weight and the drag and inertia coefficients [61,67,68]. This dirt, modeled as a shape roughness, modifies drag and lifts forces, as biological organisms could do. Many authors have reported that the presence of dirt decreases lift and increases drag. They explained that the roughness at the leading edge also had a significant influence on the lift by decreasing it and the drag by increasing it. The lift-to-drag ratio decreased with both increasing barnacle size and distribution density [69]. In [59], the potential effects of barnacles have been studied. Indeed, the lift and drag coefficients for a blade covered with idealized barnacles of different sizes and densities of distribution were determined using a wind tunnel. These lift and drag changes are made quickly until reaching a threshold (0.3 mm for a blade with a chord of 1 m) beyond which changes evolve more slowly [70]. It has been assumed that roughness or fouling presence would increase the drag coefficient up to 50% [71] (Figure 5.7). The presence of barnacles (size and density) harms the efficiency of the turbine: lift coefficients and drag [59]. In addition, all these epibiotic growths will significantly increase the weight of structures. In the case of tidal turbines, there have been a few studies, which investigate the hydrodynamic impact of biofouling [72]. To understand the phenomenon, a comparison between biofouling effects on turbines with icing effects on wind turbines will be made. Indeed, many studies have been done in icing effect with a development of many defined equations. It seems interesting to start compare these equations with other equations that could be developed and applied in the future to control the biofouling effects on speed and power generated by tidal turbines.

5.3.2   Comparison of the influence of biofouling in tidal turbines with icing in wind turbines

Mass of blade increases with the accretion of ice on wind turbine blades called mice which is directly proportional to the wind speed $V(t)$ [73,74]:

$$m_{ice}(t) = \mu V(t) \tag{5.3}$$

where μ is the mass distribution on the leading edge of the rotor blade at half the rotor radius.

Ice accretion is considered to be a linear evolution law as follows [75]:

$$m_{ice}(t) = \mu V(t)(t - t_{ice}) \tag{5.4}$$

where t_{ice} is the time when ice exposure starts. The proportionality coefficient is calculated as follows:

$$\mu = N_b \, \rho_{ice} \, k \, c_{min} + c_{max} \tag{5.5}$$

where N_b is the number of blades, ρ_{ice} is the ice density, k is a factor equal to $(0.000675 + e^{-0.32R_a})$ with R_a the radius of the turbine blade and c_{min}, c_{max} are the minimum and the maximum chord length of the blade, respectively (Figure 5.9).

The increase of rotor inertia can be calculated as

$$\Delta J(t) = \bar{J} \left[\frac{m_{ice}(t) + m}{m} - 1 \right] = \bar{J} \left[\frac{m_{ice}(t)}{m} \right] \tag{5.6}$$

where $\bar{J}$ is the nominal rotor inertia and m is the blade mass in nominal conditions. These basic equations introduced in [76] are used to derive equations applied to the impact of biofouling. They are given in the following.

The mass of the tidal turbine blade increases with the development of biofouling while being proportional to the tidal speed $V(t)$:

$$m_{biofouling}(t) = \mu V(t) \tag{5.7}$$

Similarly, we admit that biofouling accumulation is considered to follow a linear evolution law:

$$m_{biofouling}(t) = \mu V(t)(t - t_{biofouling}) \tag{5.8}$$

Where $t_{biofouling}$ is the time when biofouling coverage begin. The proportionality coefficient is calculated as follows:

$$\mu = N_b \, \rho_{biofouling} \, k \, c_{min} \, (c_{min} + c_{max}) \tag{5.9}$$

where N_b is the number of blades, $\rho_{biofouling}$ is the biofouling density, k is a factor equal to $(0.000675 + e^{-0.32R_a})$ with R_a the radius of the turbine blade and c_{min}, c_{max} the minimum and the maximum chord length of the blade, respectively (Figure 5.9).

Factors k, R_a and c_{min}, c_{max} are the same as those applied in the icing equation because μ will change only according to the biofouling density. The increase in rotor inertia can be calculated as:

$$\Delta J(t) = \bar{J} \left[\frac{m_{biofouling}(t) + m}{m} - 1 \right] = \bar{J} \left[\frac{m_{biofouling}(t)}{m} \right] \tag{5.10}$$

where $\bar{J}$ is the nominal rotor inertia and m is the blade mass in nominal conditions.

For comparison purposes between biofouling and icing, the density of the biofouling must be known. In this regard, [77] quotes as follows: Number of organisms,

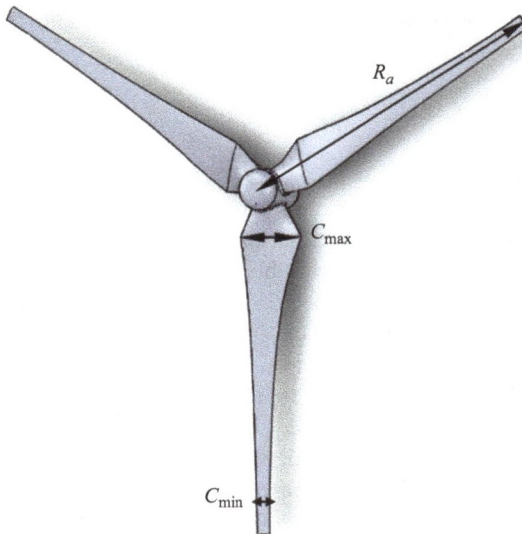

Figure 5.9   Three-blade turbine sketch

growth rate, in percentage number and percentage of area covered by different groups and biomass (g per 100 cm^2). The fouling concentration of the weekly **panels** (groups of biofouling organisms representative of a population regularly found at the end of repetitive tests) was evaluated by counting the foulants available with a magnifying lens. **Ten counts** (1 count = 1 cm^2 of area) were randomly performed on each side. However, for the monthly panels, the size of the organisms was relatively large compared to the weekly panels, so it was counted as one side of the 4-part panel and calculated accordingly. The mean density of fouling organisms observed from the short-term panels in each month (an average of the duplicate panels) was expressed as the number of individuals per 100 cm^2 except colonial forms including macroalgae. The growth rate was recorded by measuring the size of macrofoulers. Barnacles and sea anemones were measured by their diameter. Similarly, organisms such as bivalves (mussels, oysters, and other bivalves), hydroids, polychaetes were measured by length and ascidians were measured by estimating the area covered by them on the panels. Total biomass was calculated using a correction factor due to the absorption of water by the panels for specified time periods (Figure 5.10).

Besides, we have to determine the following factors: species diversity H' [78], species richness R [79,80] and evenness E [81]. They are calculated as follows to quantify and qualify the occurrence of biofouling:

$$H' = -\sum_{i}^{n} \rho_i \, ln \, \rho_i \tag{5.11}$$

where ρ_i is the proportion of individuals in species i:

$$R = \frac{S-1}{ln(N)} \tag{5.12}$$

where S and N are, respectively, the number of species and individuals in the population:

$$E = \frac{H'}{ln(S)} \tag{5.13}$$

where H' are species diversity.

Fouling is experimentally evaluated by inspection of affected materials in areas where fouling organisms are known to settle. The presence of biofouling or the biofouling community precisely is very variable in different parts of the world and also according to the seasons in the same place. Several scientific studies have confirmed it. As examples: patterns of colonization differ according to geographical location and environmental conditions, seasonal availability of colonizers' larvae, or nature of the substratum [82,83]. The extent and type of biofouling can be influenced by the depth or movement of the support [84,85]. In tropical areas, where fouling pressure may be expected to be constant throughout the year, seasonal variations in recruitment have been observed [86–88]. The sites involved in the study include the number of individuals of the main species that settle in these renewable marine energy systems. From there, we can choose the panel to study. Measuring the density of biofouling remains unreliable because density is an absolute measure of abundance and is the

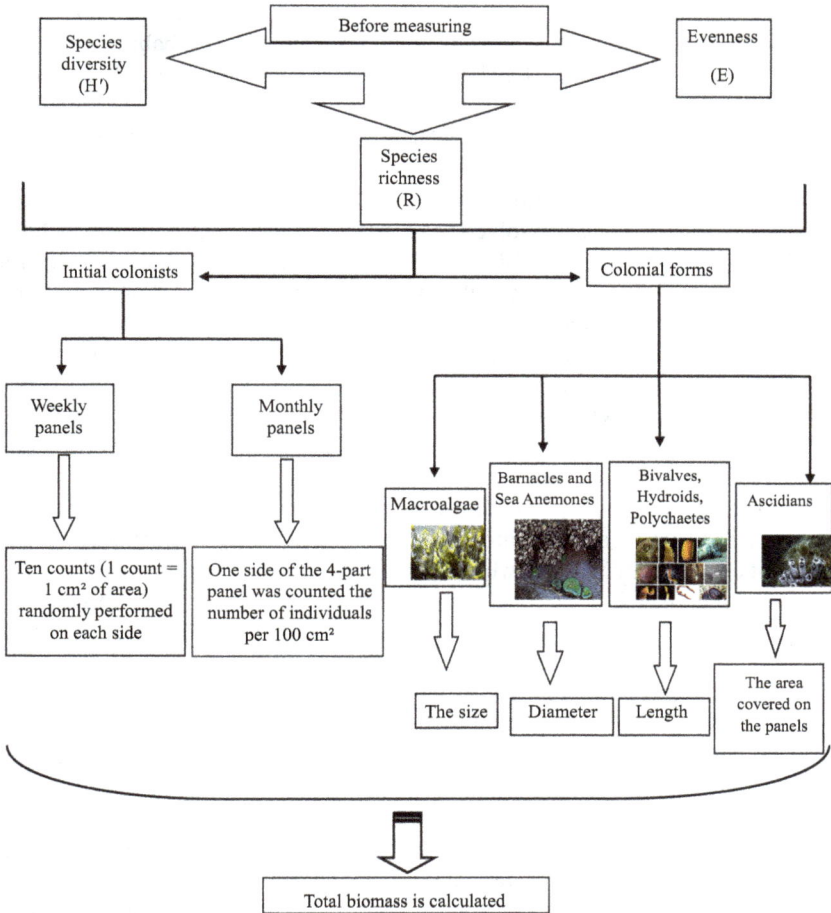

Figure 5.10   *Representative flowchart of quantitative biofouling measurement methods*

most accurate method to compare the abundance of solitary organisms. However, the density methods cannot be satisfactorily applied to species whose growth form prevents the simple definition of individuals. In addition, density is not a significant index of abundance for colonial organisms such as ascidians, sponges and Bryozoans inlaid. Additional problems with using density measurements appear in comparisons between different species. In a fouling community, a species with large, poorly distributed individuals may be more important in terms of substrate and biomass cover than a species with smaller numbers; but density measurements do not evaluate this relative importance. If a study is to focus on the settlement of a single solitary species, such as a barnacle or tubeworm, then density is the best quantitative measure [89].

Other factors are taken into account to determine this abundance as cover (the percentage of surface covered by each species [90,91]), frequency (the probability of a species occurring in a specified area [92,93]), and biomass (oven-dried panels and fouling scraped in an area in the center of each side of the panel and weighed [2,90,94–96]. All this to say that it is very difficult to get an idea about the total density of biofouling and we must take into account a case by case (the panel, the environment and all the other environmental factors: salinity, temperature water pH, as well as the dissolved elements in the water such as dissolved oxygen (DO), nitrite (NO_2), nitrate (NO_3), and other elements). For example, in [88], a study conducted in different sites in Australia led to interesting and important conclusions summarized in the following: the evaluation methods were compared on a series of panels immersed to obtain settlement data for the paint test raft in Cockburn Sound, Western Australia. A panel Queensland (North Barnard Island, dumping period 29/5/1979–28/8/1979) and one of the Naval Dockyard of Victoria (Williamstown Shipyard); immersion period from 20/04/1979 to 22/05/79) were also evaluated using the same methods as those used between sites. Table 5.1 summarizes variations in total fouling on panels from different study sites.

In summary, we can say that density is an absolute measure of abundance and is the most accurate means for comparing the abundance of solitary organisms. An estimate of coverage is sufficient when the panel covered by each species is high, such as on submerged panels for periods longer than one month, or if only total fouling coverage is required. Frequency is not an absolute measure and the results depend on the site used in the evaluation. Biomass is too heavy for the evaluation of abundant but useful individual species as a measure of total fouling abundance at both and between site studies [89]. Taking into account the information mentioned above, biodiversity is evaluated case by case and there the total biomass of biofouling can be estimated in grams per 100 cm^2 [77] or per dm^2 [89]. According to [95], recent work has investigated the relationship between proportional biofouling coverage and biofouling density using a generalized linear model with Quasi-Poisson distributions, based on a statistical hypothesis testing approach combined with Fisher's F-test [96]. The predictive relationship between percent biofouling coverage and biofouling density was reported as follows:

$$\rho_{biofouling}(t) = e^{(0.09 + 0.23H)} \tag{5.14}$$

Table 5.1   *Total fouling variation on panels from different study sites [88]*

	Western Australia	Victoria	Queensland
Density (individual/dm^2)	1,257[a]	555	25[a]
Cover (%)	12	4	97
Frequency (%)	100	88	100
Biomass (g/dm^2)	0.53	0.06	0.35

[a]Excluding algae and/or hydroids.

where ρ is the biofouling density (g cm^{-3}) and H is the proportional biofouling coverage (a value between 0 and 1). Taking all these factors and elements into account, the biofouling density is calculated to be included in (5.3) and becomes:

$$\mu = N_b \, \rho_{biofouling} \, k \, c_{\min} \, (c_{\min} + c_{\max})$$
$$= N_b \, e^{(0.09+0.23H)} \, k \, c_{\min} \, (c_{\min} + c_{\max}) \tag{5.15}$$

The above-proposed methodology for the study of biofouling density is very specific and well detailed. Nevertheless, it requires a lot of time and resources. The purpose of this study is to build a simulation model of biofouling "density" in tidal turbines that will be dedicated and applied to a particular site.

5.4 Biofouling impact on tidal stream turbines hydrodynamic and electric performance

In terms of hydrodynamics, it is important to study how biological fouling will impact on energy absorption in tidal turbines as well as the electrical power generated. Indeed, biofouling reduces the efficiency of the turbine blades and thus decreases the overall energy generation. Regarding blade fouling, a significant decrease in power at higher tip speed ratio was noticed [82]. In the same way as for the previous hydrodynamic part where (5.10) was highlighted, a comparison with it will also give a little more clarification about the influence of biofouling on the power coefficient. The hydrodynamic (mechanical) power extracted by the tidal turbine is expressed according to the following equation:

$$P_a = \frac{1}{2} D \pi r^2 C_p \, (\lambda, \beta) \, V^3(t) \tag{5.16}$$

where V is the tidal speed, D is the water density, r is the tidal turbine rotor radius, and C_P is the power coefficient representing the conversion efficiency. This coefficient depends on both the blade pitch angle and the tip speed ratio [76]. In [71], it has been clearly shown that the tip speed ratio at maximum C_P was shifted upward with increasing roughness, which will automatically affect the power P_a in (5.14) by decreasing it.

As in [76], the tidal turbine hydrodynamic torque is given by

$$T_{hydro}(t) = \frac{P_a(t)}{\omega(t)} = \frac{D \pi r^3 C_p \, (\lambda, \beta) \, V^2}{2\lambda} \tag{5.17}$$

where

$$\lambda = \frac{\omega(t)}{V(t)} \tag{5.18}$$

ω being the angular shaft speed of the turbine.

According to [83], the following equation is proposed to model the power coefficient:

$$C_p\,(\lambda, \beta) = C_1 \left(\frac{k_1}{\lambda} + K_2 + \beta + K_3 \right) e^{\frac{K_4}{\lambda}} \qquad (5.19)$$

Coefficients C_1, K_1, K_2, K_3, and K_4 depend on the shape of the blade and its hydro-dynamic performance [83,84]. When C_p is controlled at the maximum value, the maximum mechanical power is extracted from the tidal energy. In [76], a generator torque control with constant blade pitch to maximize the energy capture of a variable speed tidal turbine is considered to operate with low and medium tidal speed.

The mechanical equation governing the turbine can be simplified as follows:

$$\bar{J}\frac{d\omega}{dt} = -B\omega(t) + T_{hydro}(t) - T_e(t) \qquad (5.20)$$

where $\bar{J}$ is the rotor nominal inertia, B is the viscous friction coefficient, T_{hydro} is the hydrodynamic torque, T_e is the generator electromagnetic torque, and ω is the turbine angular shaft speed.

The electromagnetic torque, which is the factor sought to understand the influence of biofouling, will be given by

$$T_e(t) = T_{hydro}(t) - J\frac{d\omega}{dt} - B\omega(t) \qquad (5.21)$$

The impact of B is evaluated using the viscous friction forces, which depend on several factors, including fluid velocity (deep sea currents) that will not exceed 20 m/s [97]. In this context, these forces are calculated according to the following Stocks equations [98]:

• At very low speed ($V < 5$ m/s)

$$F_{friction} = -k\eta V \qquad (5.22)$$

where k is the characteristic coefficient of the solid geometry (it reflects the biofouling geometry); $k = 6\pi R$, is the fluid viscosity coefficient, R is the radius of the geometry, and V is the estimated speed.

• At higher speed ($5 < V < 20$ m/s)

$$F_{friction} = -C_x\,\frac{1}{2}\,D\,V^2\,S\,\hat{V} \qquad (5.23)$$

where D is the density of the fluid; S is the solid area which has a perpendicular direction to the speed and area here represents area of biofouling particles; C_x is the drag coefficient characterizing the geometry (Figure 5.11) of the biofouling (without unity) and $\hat{V} = \sqrt{v^2}$ (the root mean square velocity is the square root of the average of the square velocity).

The general expression for the viscous friction coefficient B is given by the following equation [99]:

$$B = |\frac{F_{friction}}{V}| \qquad (5.24)$$

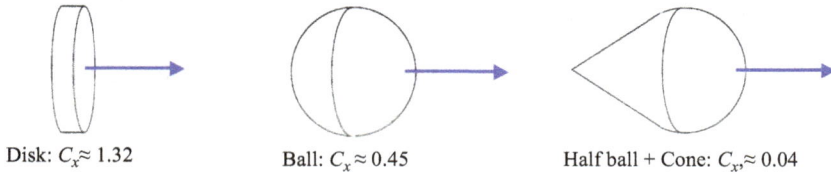

Disk: $C_x \approx 1.32$ Ball: $C_x \approx 0.45$ Half ball + Cone: $C_x \approx 0.04$

Figure 5.11   Drag coefficient according to the biofouling geometry

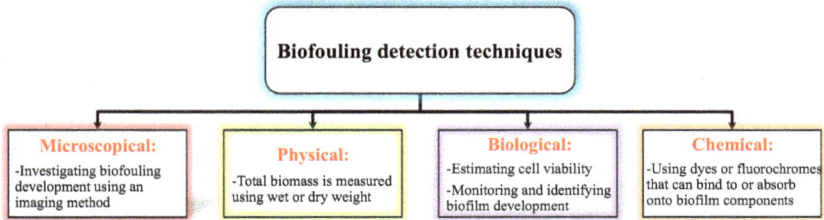

Figure 5.12   Biofouling detection techniques summary [100]

From the above-given equations, it is clearly shown that the viscous friction coefficient increases with the friction force. All these elements will allow the evaluation of the power and electric torque generated in presence of biofouling (5.19). This will therefore be a first modeling approach allowing simulation studies related to biofouling.

5.4.1 Biofouling detection techniques briefly

There is no universal agreement on the best methods to detect biofouling [100]. The various biofouling detection methods were divided into four groups by Azeredo *et al.* [100]: chemical, physical, microscopical, and biological. A short definition of these techniques is summarized in Figure 5.12.

In addition to these four classifications, the following characteristics of biofilm detection methods can be used for definition and classification purposes: online, in-person, real-time, non-destructive, representative, accurate, and repeatable. Automatic monitoring methods are also available [101]. Additionally, outcomes of the various biofilm detection methods can vary. Microbial activity, total cell counts, 2D bacterial distribution in the biofilm, 3D biofilm structure, and the ability to distinguish between different biofilm components are all outcomes that could be attained by using various detection techniques [102].

5.5   Conclusions and perspectives

Tidal turbines, underwater turbines that transform marine currents into electricity, have for many years been the source of hope for a clean energy source, without the visual nuisance caused by wind turbines. In June 2018, the Naval Group subsidiary inaugurated the world's first wind turbine assembly factory in Cherbourg. Recall that the French company Sabella has installed a tidal turbine offshore of Ouessant Island.

References

[1]   Sagol E, Reggio M, and Ilinca A. Issues concerning roughness on wind turbine blades. *Renewable and Sustainable Energy Reviews*. 2013;23:514–525.

[2]   Batten W, Bahaj A, Molland A, *et al.* The prediction of the hydrodynamic performance of marine current turbines. *Renewable Energy*. 2008;33(5):1085–1096.

[3]   Lewis M, Neill S, Robins P, *et al.* Resource assessment for future generations of tidal-stream energy arrays. *Energy*. 2015;83:403–415.

[4]   Myers L and Bahaj A. Experimental analysis of the flow field around horizontal axis tidal turbines by use of scale mesh disk rotor simulators. *Ocean Engineering*. 2010;37(2–3):218–227.

[5]   Neill SP, Hashemi MR, and Lewis MJ. Tidal energy leasing and tidal phasing. *Renewable Energy*. 2016;85:580–587.

[6]   Gehrke T and Sand W. Interactions between microorganisms and physiochemical factors cause MIC of steel pilings in harbors (ALWC). In: *CORROSION 2003*. OnePetro; 2003.

[7]   Titah-Benbouzid H and Benbouzid M. Biofouling issue on marine renewable energy converters: a state of the art review on impacts and prevention. *International Journal of Energy Conversation*. 2017;5(3):67.

[8]   Bahaj A, Batten W, and McCann G. Experimental verifications of numerical predictions for the hydrodynamic performance of horizontal axis marine current turbines. *Renewable Energy*. 2007;32(15):2479–2490.

[9]   Magana D, Monfardini D, and Uihlein A. Technology market and economic aspects of ocean energy in Europe. RC Ocean Energy Status Report. 2016. p. 4–10.

[10]   Hydro B and Hydro B. Green Energy Study For British Columbia, Phase 2: Mainland. BC Hydro, Green & Alternative Energy Division; 2002.

[11]   Blunden L and Bahaj A. Tidal energy resource assessment for tidal stream generators. *Proceedings of the Institution of Mechanical Engineers, Part A: Journal of Power and Energy*. 2007;221(2):137–146.

[12]   Bahaj A and Myers L. Analytical estimates of the energy yield potential from the Alderney Race (Channel Islands) using marine current energy converters. *Renewable Energy*. 2004;29(12):1931–1945.

[13]   Myers L and Bahaj A. Simulated electrical power potential harnessed by marine current turbine arrays in the Alderney Race. *Renewable Energy*. 2005;30(11):1713–1731.

[14]   Ng KW, Lam WH, and Ng KC. 2002–2012: 10 years of research progress in horizontal-axis marine current turbines. *Energies*. 2013;6(3):1497–1526.

[15] Bahaj A, Molland A, Chaplin J, *et al*. Power and thrust measurements of marine current turbines under various hydrodynamic flow conditions in a cavitation tunnel and a towing tank. *Renewable Energy*. 2007;32(3): 407–426.

[16] Batten W, Bahaj A, Molland A, *et al*. Experimentally validated numerical method for the hydrodynamic design of horizontal axis tidal turbines. *Ocean Engineering*. 2007;34(7):1013–1020.

[17] Luznik L, Flack KA, Lust EE, *et al*. The effect of surface waves on the performance characteristics of a model tidal turbine. *Renewable Energy*. 2013;58:108–114.

[18] Ocean Energy Systems Annual Report 2010, Implementing Agreement on Ocean Energy Systems; 2010 [cited 2012 October 17]. http://www.ocean-energy-systems.org/library/annual_reports/.

[19] Ocean Energy Systems Annual Report 2010, Implementing Agreement on Ocean Energy Systems; 2011 [cited 2012 October 17]. http://www.ocean-energy-systems.org/library/annual_reports/.

[20] Union E. Directive 2009/28/EC of the European Parliament and of the Council of 23 April 2009 on the promotion of the use of energy from renewable sources and amending and subsequently repealing Directives 2001/77/EC and 2003/30/EC. *Official Journal of the European Union*. 2009;5:2009.

[21] 4coffshore Global Wind Farm; [cited 2012 January 30]. http://www. 4coffshore.com/offshorewind/.

[22] Notre-planete.info; [cited 2022 March 07]. https://www.notre-planete.info/actualites/4022-hydrolienne-Bretagne-test?nd=10#com/.

[23] L'usine Nouvelle; [cited 2018 June 18]. https://www.usinenouvelle.com/article/openhydro-et-edf-arretent-leur-projet-d-hydroliennes-au-large-de-pai mpol-brehat.N611153/.

[24] Renewable Green Energy Power; [cited 2018 June 24]. http://www. renewablegreenenergypower.com/tidal- energy-tidal-power-facts/.

[25] Bernardo C. The fouling of catalysts by deposition of filamentous carbon. In: *Fouling Science and Technology*. Springer, New York, NY; 1988. p. 369–389.

[26] Fingerman M. *Endocrinology and Reproduction: Recent Advances in Marine Biotechnology*. CRC Press, London; 1997.

[27] Railkin AI. *Marine Biofouling: Colonization Processes and Defenses*. CRC Press, London; 2003.

[28] Hellio C and Yebra D. *Advances in Marine Antifouling Coatings and Technologies*. Elsevier, New York, NY; 2009.

[29] Copisarow M. Marine fouling and its prevention. *Science*. 1945;101(2625): 406–407.

[30] Institution WHO of Ships USNDB. Marine Fouling and Its Prevention. 580. United States Naval Institute; 1952.

[31] Barnes H and Ray D. Marine boring and fouling organisms. *AIBS Bulletin*. 1960;10:41.

[32] Callow ME and Callow JA. Marine biofouling: a sticky problem. *Biologist*. 2002;49(1):1–5.

[33]   Biofouling formation and remedial measures; [cited 2004 March 12]. http://www.parliament.vic.gov.au/enrc/default.html//.

[34]   Callow JA and Callow ME. Trends in the development of environmentally friendly fouling-resistant marine coatings. *Nature Communications.* 2011;2(1):1–10.

[35]   Abbott A, Abel P, Arnold D, *et al.* Cost–benefit analysis of the use of TBT: the case for a treatment approach. *Science of the Total Environment.* 2000;258 (1–2):5–19.

[36]   Rouhi AM. The squeeze on tributyltins. *Chemical & Engineering News.* 1998;76(17):41–42.

[37]   Gabbay J, Borkow G, Mishal J, *et al.* Copper oxide impregnated textiles with potent biocidal activities. *Journal of Industrial Textiles.* 2006;35(4): 323–335.

[38]   Brancato MS and MacLellan D. Impacts of invasive species introduced through the shipping industry. In: *Oceans' 99. MTS/IEEE. Riding the Crest into the 21st Century. Conference and Exhibition. Conference Proceedings* (IEEE Cat. No. 99CH37008). vol. 2. IEEE; 1999. p. 676–vol.

[39]   Reise K, Gollasch S, and Wolff WJ. Introduced marine species of the North Sea coasts. *Helgoländer Meeresuntersuchungen.* 1998;52(3):219–234.

[40]   Mineur F, Cook EJ, Minchin D, *et al.* Changing coasts: marine aliens and arti cial structures. In: *Oceanography and Marine Biology.* CRC Press, London; 2012. p. 198–243.

[41]   Polagye B and Thomson J. *Screening for Biofouling and Corrosion of Tidal Energy Device Materials: In-Situ Results for Admiralty Inlet, Puget Sound,* Washington; 2010.

[42]   Wahl M. Marine epibiosis. I. Fouling and antifouling: some basic aspects. *Marine Ecology Progress Series.* 1989;58:175–189.

[43]   Achinas S, Charalampogiannis N, and Euverink GJW. A brief recap of microbial adhesion and biofilms. *Applied Sciences.* 2019;9(14):2801.

[44]   Alav I, Sutton JM, and Rahman KM. Role of bacterial efflux pumps in biofilm formation. *Journal of Antimicrobial Chemotherapy.* 2018;73(8):2003–2020.

[45]   Santos ALSd, Galdino ACM, Mello TPd, *et al. What are the Advantages of Living in a Community? A Microbial Biofilm Perspective!* Memórias do Instituto Oswaldo Cruz. 2018. p. 113.

[46]   Garrett TR, Bhakoo M, and Zhang Z. Bacterial adhesion and biofilms on surfaces. *Progress in Natural Science.* 2008;18(9):1049–1056.

[47]   Jones G. The battle against marine biofouling: a historical review. In: *Advances in Marine Antifouling Coatings and Technologies.* 2009. p. 19–45.

[48]   Griebe T and Flemming HC. *Rotating Annular Reactors for Controlled Growth of Biofilms.* Technomic Publishing Co, Inc., Lancaster, PA; 2000.

[49]   Ralston E and Swain G. Bioinspiration—the solution for biofouling control? *Bioinspiration & Biomimetics.* 2009;4(1):015007.

[50]   Callow ME, Pitchers R, and Milne A. The control of fouling by non-biocidal systems. In: *Studies in Environmental Science,* vol. 28. Elsevier, New York, NY; 1986. p. 145–158.

[51] Bhushan B. Adhesion and stiction: mechanisms, measurement techniques, and methods for reduction. *Journal of Vacuum Science & Technology B: Microelectronics and Nanometer Structures Processing, Measurement, and Phenomena*. 2003;21(6):2262–2296.

[52] Sheng X, Ting YP, and Pehkonen SO. Force measurements of bacterial adhesion on metals using a cell probe atomic force microscope. *Journal of Colloid and Interface Science*. 2007;310(2):661–669.

[53] Landoulsi J, Cooksey K, and Dup res V. Review—Interactions between diatoms and stainless steel: focus on biofouling and biocorrosion. *Biofouling*. 2011;27(10):1105–1124.

[54] Nguyen T, Roddick FA, and Fan L. Biofouling of water treatment membranes: a review of the underlying causes, monitoring techniques and control measures. *Membranes*. 2012;2(4):804–840.

[55] Vanysacker L, Boerjan B, Declerck P, *et al*. Biofouling ecology as a means to better understand membrane biofouling. *Applied Microbiology and Biotechnology*. 2014;98(19):8047–8072.

[56] Grishkin V, Iakushkin O, and Stepenko N. Biofouling detection based on image processing technique. In: *2017 Computer Science and Information Technologies (CSIT)*. IEEE; 2017. p. 158–161.

[57] Fey PD and Olson ME. Current concepts in biofilm formation of *Staphylococcus epidermidis*. *Future Microbiology*. 2010;5(6): 917–933.

[58] Flemming HC. Physico-chemical properties of biofilms—a short review. *Biofilms in the Aquatic Environment*. Royal Society of Chemistry, Cambridge; 1999.

[59] Orme J, Masters I, and Griffiths R. Investigation of the effect of biofouling on the efficiency of marine current turbines. In: *Proceedings of the MAREC*; 2001. p. 91–99.

[60] Khor YS and Xiao Q. CFD simulations of the effects of fouling and antifouling. *Ocean Engineering*. 2011;38(10):1065–1079.

[61] Racerocks. 6-Month Fouling Records; [cited 2018 June 12]. http://www.racerocks.com/racerock/energy/tidalenergy/april07fouling/fouling.htm/.

[62] Thomason J, Hills J, Clare A, *et al*. Hydrodynamic consequences of barnacle colonization. *Hydrobiologia*. 1998;375:191–201.

[63] Koehl M. Mini review: hydrodynamics of larval settlement into fouling communities. *Biofouling*. 2007;23(5):357–368.

[64] Demirel YK, Uzun D, Zhang Y, *et al*. Effect of barnacle fouling on ship resistance and powering. *Biofouling*. 2017;33(10): 819–834.

[65] Hunt MJ and Alexander C. Feeding mechanisms in the barnacle Tetraclita squamosa (Bruguiere). *Journal of Experimental Marine Biology and Ecology*. 1991;154(1):1–28.

[66] Schultz MP, Kavanagh CJ, and Swain GW. Hydrodynamic forces on barnacles: implications on detachment from fouling-release surfaces. *Biofouling*. 1999;13(4):323–335.

[67]   Theophanatos A and Wolfram J. Hydrodynamic loading on macro-roughened cylinders of various aspect ratios. *Journal of Offshore Mechanics and Arctic Engineering – Transactions of the Asme*. 1989;111:214–222.

[68]   Shi W, Park HC, Baek JH, *et al*. Study on the marine growth effect on the dynamic response of offshore wind turbines. *International Journal of Precision Engineering and Manufacturing*. 2012;13(7):1167–1176.

[69]   Janiszewska J, Ramsay RR, Hoffmann M, *et al*. *Effects of Grit Roughness and Pitch Oscillations on the S814 Airfoil*. National Renewable Energy Lab. (NREL), Golden, CO; 1996.

[70]   Ren N and Ou J. Numerical simulation of surface roughness effect on wind turbine thick airfoils. In: *2009 Asia-Pacific Power and Energy Engineering Conference*. IEEE; 2009. p. 1–4.

[71]   Walker JM, Flack KA, Lust EE, *et al*. Experimental and numerical studies of blade roughness and fouling on marine current turbine performance. *Renewable Energy*. 2014;66:257–267.

[72]   Day A, Babarit A, Fontaine A, *et al*. Hydrodynamic modelling of marine renewable energy devices: a state of the art review. *Ocean Engineering*. 2015;108:46–69.

[73]   Frohboese P and Anders A. Effects of icing on wind turbine fatigue loads. *Journal of Physics: Conference Series*. 2007;75:012061.

[74]   Myers TG. Extension to the Messinger model for aircrafticing. *AIAA Journal*. 2001;39(2):211–218.

[75]   Saleh S, Ahshan R, and Moloney C. Wavelet-based signal processing method for detecting ice accretion on wind turbines. *IEEE Transactions on Sustainable Energy*. 2012;3(3):585–597.

[76]   Corradini ML, Ippoliti G, and Orlando G. A robust observer for detection and estimation of icing in wind turbines. In: *IECON 2016—42nd Annual Conference of the IEEE Industrial Electronics Society*. IEEE, New York, NY; 2016. p. 1894–1899.

[77]   Shannon CE. A mathematical theory of communication. *The Bell System Technical Journal*. 1948;27(3):379–423.

[78]   Gleason HA. On the relation between species and area. *Ecology*. 1922;3(2):158–162.

[79]   Margalef R. Temporal succession and spatial heterogeneity in natural phytoplankton. In: *Symposium held at Scripps Institution of Oceanography*, 24 March–2 April 1956, California, 1958.

[80]   Pielou EC. The measurement of diversity in different types of biological collections. *Journal of Theoretical Biology*. 1966;13:131–144.

[81]   Gribben PE, Marshall DJ, and Steinberg PD. Less inhibited with age? Larval age modifies responses to natural settlement inhibitors. *Biofouling*. 2006;22(02):101–106.

[82]   Underwood A and Chapman M. Early development of subtidal macrofaunal assemblages: relationships to period and timing of colonization. *Journal of Experimental Marine Biology and Ecology*. 2006;330(1): 221–233.

[83] Glasby T. Development of sessile marine assemblages on fixed versus moving substrata. *Marine Ecology Progress Series*. 2001;215:37–47.

[84] Satheesh S and Wesley SG. Seasonal variability in the recruitment of macro-fouling community in Kudankulam waters, east coast of India. *Estuarine, Coastal and Shelf Science*. 2008;79(3):518–524.

[85] Dharmaraj S and Chellam A. Settlement and growth of barnacle and associated fouling organisms in pearl culture farm in the Gulf of Mannar. In: *Proceedings of the Symposium on Coastal Aquaculture, Part 2, MBAI*, 12–18 January 1980.

[86] Leca L. Etude des épibiontes associéss à l'huître perlière Pinctada margaritifera dans deux atolls de Polynésie Française. Thèse de doctorat. Université de la Polynésie Française; 1992.

[87] Rodriguez LF and Ibarra-Obando SE. Cover and colonization of commercial oyster (Crassostrea gigas) shells by fouling organisms in San Quintin Bay, Mexico. *Journal of Shellfish Research*. 2008;27(2):337–343.

[88] Lewis JA. *A Comparison of Possible Methods for Marine Fouling Assessment during Raft Trails*. Materials Research Labs ASCOT VALE (Australia); 1981.

[89] Sutherland JP and Karlson RH. Development and stability of the fouling community at Beaufort, North Carolina. *Ecological Monographs*. 1977;47(4):425–446.

[90] Russ GR. Effects of predation by fishes, competition, and structural complexity of the substratum on the establishment of a marine epifaunal community. *Journal of Experimental Marine Biology and Ecology*. 1980;42(1): 55–69.

[91] Grovhoug JG. Marine environmental assessment at three sites in Pearl Harbor, Oahu, August–October 1978. Final Report August–October 1978. Naval Ocean Systems Center, San Diego, CA; 1979.

[92] Rastetter E and Cooke W. Responses of marine fouling communities to sewage abatement in Kaneohe Bay, Oahu, Hawaii. *Marine Biology*. 1979;53(3): 271–280.

[93] DePalma JR. *Marine Biofouling Studies in the Gulf of Siam*. Naval Oceanographic Office, Washington, DC; 1977.

[94] Lewis JA. *Settlement of Fouling Organisms at the JTTRE North Barnard Island Raft Site*. Materials Research Labs Ascot Vale (Australia); 1981.

[95] Macleod AK, Stanley MS, Day JG, *et al.* Biofouling community composition across a range of environmental conditions and geographical locations suitable for floating marine renewable energy generation. *Biofouling*. 2016;32(3):261–276.

[96] Zuur AF, Ieno EN, Walker NJ, *et al. Mixed Effects Models and Extensions in Ecology with R*, vol. 574. Springer, New York, NY; 2009.

[97] Benelghali S. *On Multiphysics Modeling and Control of Marine Current Turbine Systems*. Université de Bretagne occidentale-Brest; 2009.

[98] Temam R. *Navier–Stokes Equations: Theory and Numerical Analysis*, vol. 343. American Mathematical Soc.; 2001.

[99]   Wiegel F and Mijnlieff P. Intrinsic viscosity and friction coefficient of permeable macromolecules in solution. *Physica A: Statistical Mechanics and its Applications*. 1977;89(2):385–396.

[100]   Azeredo J, Azevedo NF, Briandet R, *et al.* Critical review on biofilm methods. *Critical Reviews in Microbiology*. 2017;43(3):313–351.

[101]   Klahre J and Flemming H. Monitoring of biofouling in papermill process waters. *Water Research*. 2000;34(14):3657–3665.

[102]   Turan O, Demirel YK, Day S, *et al.* Experimental determination of added hydrodynamic resistance caused by marine biofouling on ships. *Transportation Research Procedia*. 2016;14:1649–1658.

Conclusion

Mohamed Benbouzid[1] and Demba Diallo[2]

In summary, this book presented an overview of the energy conversion drivetrain design option, control, resilience (fault–tolerant control), and monitoring including the specific issue of biofouling. It provided methodologies and algorithms with several illustrative examples and practical case studies while highlighting some prospective investigations. It includes extensive application features not found in academic text-books and it can be considered a potential guide for prospective tidal stream turbine developers.

Further investigations are needed to deal with tidal stream turbines-specific challenges and increase their competitiveness and deployment. In particular, opportunities were identified to address critical maintenance and survivability issues: biofouling detection, estimation, and resilience to failure. This objective may be particularly relevant to the marine renewable energy industry, as maintenance in a high-energy environment is costly and time-consuming.

In this context, biofouling is expected to be monitored using electrical measurements (mainly the current flowing in the windings of the electrical generator: doubly-fed induction or permanent magnet synchronous generators). Indeed, the main effects of biofouling on a tidal turbine are reduced energy production, reduced turbine durability, physical deformation of the turbine blades, significant turbine shaft eccentricity, and bearing damage. These effects will obviously impact the turbine electric generator current. Early detection and assessment of biofouling is therefore essential to plan for early removal of the fouling before it accumulates. For monitoring purposes, first-order modeling of the biofouling is carried out (Chapter 4) for the design of specific advanced signal processing techniques of the generator current for biofouling detection. Afterward, it is expected to estimate the detected biofouling initiation using either a failure (i.e., biofouling) severity index or designing a specific observer of the generator rotor angular acceleration to estimate the inertia, which is impacted by the biofouling. Finally, it is also expected to predict (prognosis) biofouling evolution for maintenance scheduling. In this context, machine learning techniques should be privileged. Indeed, they are becoming an undeniable alternative to taking advantage of the increase in available data. They allow better modeling (knowledge) of this complex and multi-physics system. Hybrid approaches using pre-processed signals (i.e., generator current) and machine learning-based strategies have shown potential for effective prognosis.

[1]University of Brest, CNRS, Institut de Recherche Dupuy de Lôme, France
[2]Université Paris-Saclay, CentraleSupelec, CNRS, Group of Electrical Engineering Paris, France

Index